T0245308

Analytical Chemistry for Assessing Medication Adherence

Emerging Issues in Analytical Chemistry

Series Editor
Brian F. Thomas

AMSTERDAM · BOSTON · HEIDELBERG · LONDON
NEW YORK · OXFORD · PARIS · SAN DIEGO
SAN FRANCISCO · SINGAPORE · SYDNEY · TOKYO

Analytical Chemistry for Assessing Medication Adherence

Sangeeta Tanna
De Montfort University, Leicester, United Kingdom

Graham Lawson
De Montfort University, Leicester, United Kingdom

ELSEVIER

AMSTERDAM • BOSTON • HEIDELBERG • LONDON
NEW YORK • OXFORD • PARIS • SAN DIEGO
SAN FRANCISCO • SINGAPORE • SYDNEY • TOKYO

Elsevier
Radarweg 29, PO Box 211, 1000 AE Amsterdam, Netherlands
The Boulevard, Langford Lane, Kidlington, Oxford OX5 1GB, UK
50 Hampshire Street, 5th Floor, Cambridge, MA 02139, USA

Published in cooperation with RTI Press at RTI International, an independent, nonprofit research institute
that provides research, development, and technical services to government and commercial clients worldwide
(www.rti.org). RTI Press is RTI's open-access, peer-reviewed publishing channel. RTI International is a trade
name of Research Triangle Institute.

British Library Cataloguing-in-Publication Data
A catalogue record for this book is available from the British Library

Library of Congress Cataloging-in-Publication Data
A catalog record for this book is available from the Library of Congress

ISBN: 978-0-12-805463-5

For Information on all Elsevier publications
visit our website at http://www.elsevier.com/

Working together
to grow libraries in
developing countries

www.elsevier.com • www.bookaid.org

Publisher: John Fedor
Acquisition Editor: Kathryn Morrissey
Editorial Project Manager: Jill Cetel
Production Project Manager: Vijayaraj Purushothaman
Designer: Matthew Limbert

Typeset by MPS Limited, Chennai, India

CONTENTS

BIOGRAPHIES

Sangeeta Tanna

Sangeeta Tanna has a PhD in Pharmaceutics and is a Reader in Pharmaceutical Bioanalysis. Her expertise and research interests are the bioanalysis and drug delivery fields. Her early research focused on developing instrumental methods for measuring the glucose-dependent delivery of insulin from a "smart" gel. This led to the development of microanalytical methodologies for the determination of therapeutic drugs from dried blood spots based on LC-MS, LC-MS/MS, and LC-HRMS studies. Applications of this work to patient healthcare include improved medications for newborn babies and to assess adherence to cardiovascular drug therapy for adult patients. She has a track record of published papers, invited international conference presentations, grants, and awards for this research.

Graham Lawson

Graham Lawson has a PhD in Analytical Chemistry and has expertise in instrumental analysis, in disparate areas such as environmental exposure in the polymer industry, the identification of migrants from food packaging, and factors influencing drug delivery and clinical applications. The unifying themes are the detection of ultra-low levels of contamination in complex matrices and the protection of people from adverse exposures. He was co-opted onto a NATO special studies group on the standoff detection of radiation. His current research interests include novel analytical techniques applied to dried blood spot analyses and to counterfeit drug detection.

This is a book about a problem, a worldwide problem, a problem that needs the input from a disparate range of skills for a solution. The problem is poor adherence to prescription medicines. The challenge for the authors was to communicate the extent of the problem and the efforts to overcome it in a way that is informative to everybody with an interest in this situation.

In this book, the authors have attempted to capture a global perspective of the problem of nonadherence and to apply their bioanalytical and instrumental analytical chemistry expertise to considering alternative objective but yet patient-friendly methods to assess individual adherence. This approach has proved to be a bigger challenge than anticipated with a rapidly growing worldwide interest in this problem becoming evident from public domain but not necessarily peer-reviewed literature.

It is a simple fact that at least half of all prescription medicines may be wasted because they are not taken as prescribed. Under these circumstances, patients get no benefit from their prescribed drug therapy and billions of dollars are lost to the healthcare provider. In the cold light of day, this seems an unrealistic situation—why should patients suffer unnecessarily, why throw away such large sums of money? And yet it happens. The pharmaceutical industry spends billions of dollars on marketing its products but there are few examples of it helping people to take them correctly. Healthcare professionals carry out patient-centered checks on adherence, but in many cases, now accept that the results will be optimistic. Patients present various reasons for not taking their prescribed medicines, many are understandable, even more are not, but in most cases their clinicians are unaware of their actions.

Prescriptions are meant to achieve therapeutic levels of a particular drug in a patient's bloodstream. Analytical chemistry now has the techniques to unequivocally measure these drug levels. Direct analysis of blood or other biosamples can confirm that the clinician's prescription has achieved these levels, indicate patient adherence and identify any unexpected drug–drug interaction or even an incorrect diagnosis. This level of information is necessary because the therapeutic window for some drugs is so narrow that if it is not achieved patient lives may be at risk.

The book follows a logical sequence from an assessment of the global problem of medicine nonadherence, in the chapter "Medication Adherence," to considering how noninvasive samples can be obtained from a patient to assess the relevant drug levels. This approach, outlined in the chapter "Opportunities and Challenges for Analytical Chemistry in the Assessment of Medication Adherence," requires knowledge of the factors affecting drug disposition in the body and identifies both possible suitable biosamples and the anticipated drug levels to be detected. Detection of the target drug presents analytical chemistry with two challenges, namely specificity and sensitivity. Biosamples are complex mixtures, and the chapter "Analytical Chemistry Methods for the Assessment of Medication Adherence" details the different approaches taken to provide the specificity to ensure that only the target species will be measured. Developing sufficient sensitivity is a combination of sample size, drug concentration, and possibly sample pre-work-up. Factors affecting sample throughput and some notional instrumental comparisons are documented. The alternative approach of measuring a disease symptom, as a measure of adherence, is also considered. The complementary nature of both approaches is indicated from results for different illnesses.

The authors accept that the chapter "Application of Bioanalytical Methods to Assess Medication Adherence in Clinical Settings" probably underrepresents all the work that is done worldwide to measure patient sample drug levels and hence assess adherence to medication. The missing data will be shared between hospital laboratories and private health laboratories and is secured by patient confidentiality. Nevertheless, there is sufficient material presented in the chapter "Application of Bioanalytical Methods to Assess Medication Adherence in Clinical Settings" to demonstrate both international interest and collaboration in the direct assessment of medication adherence in a wide range of illnesses. The chapter "Medication Adherence: Where Do We Go From Here?" attempts to bring together the problems perceived by different stakeholders in the provision of the analytical services necessary to deliver a more objective assessment of adherence and suggests some possible routes forward including the concept of the attained drug level being a measure of adherence rather than a determination of "when" or "if" the medication was taken.

The authors hope that this book will be of interest to a broad readership audience including healthcare professionals; biomedical scientists; analytical instrument manufacturers and users; sectors of the pharmaceutical industry; pharmacy, medical and biomedical students, and academics; researchers in drug analysis and related clinical chemistry disciplines; regulatory bodies; government departments and policymakers for public health issues; patients and journalists with an interest in public health issues; international public health agencies such as the World Health Organization.

ACKNOWLEDGMENTS

We gratefully acknowledge Steve Lawson for production of most of the images. We thank Carolyn Jones and Dr Elaine Woodland for proof-reading the manuscript and for their valuable suggestions, without which the manuscript would have been the poorer. Thank you Brian for providing the opportunity and challenge to commit our thoughts and ideas to paper.

Sangeeta Tanna

Graham Lawson

CHAPTER 1

Medication Adherence

INTRODUCTION

Medicines are the most common intervention in healthcare and are crucial in maintaining health, preventing illness, managing chronic conditions, and curing disease. Getting the most from medicines for both patients and health service providers is becoming increasingly important as people are living longer and are suffering from more than one long-term condition. Medicines are dispensed with the expectation that they will be taken exactly as prescribed. However, their benefits are often not realized because, alarmingly, only 50% of medicines prescribed in developed countries are taken as recommended and this figure is estimated to be lower in developing countries.[1] In the words of Dr C. Everett Koop, "drugs don't work in patients who don't take them."[2] This lack of adherence to prescribed therapies is termed "medication nonadherence" and is one of the most understated problems facing healthcare systems worldwide.

WHAT IS MEDICATION ADHERENCE?

Adherence to (or compliance with) a medication regimen is generally defined as the ability of a patient to take their medications in the way recommended by their healthcare providers.[1,2]

The terms adherence and compliance are commonly used interchangeably to describe the extent to which a patient takes their medication as prescribed.[3] *Compliance* is defined as the extent to which the patient's behavior matches the prescriber's recommendations. *Adherence* in turn is the extent to which the patient's behavior matches the agreed recommendations from the prescriber. Their meanings are, therefore, somewhat different since adherence presumes the patient's agreement with the recommendations, whereas compliance suggests that a patient is passively following a doctor's instructions, rather than actively collaborating in the treatment process.[3] Adherence

Analytical Chemistry for Assessing Medication Adherence. DOI: http://dx.doi.org/10.1016/B978-0-12-805463-5.00001-8

is nowadays preferred by many to compliance because of its emphasis on the need for agreement[4] and because healthcare communities are adopting concepts of patient-centered care and activation.[5] Regardless of which term is used, it is clear that the medications will benefit patients only if they follow prescribed treatment regimens correctly.

Taking medications correctly also involves more than simply reading the "directions on the bottle." Proper adherence to a medication regimen involves six key factors:[6]

- taking the right medication
- taking the medication at the right dose
- taking the medication at the right time
- following the right schedule
- taking the medication under the right conditions, for example, medication should be taken on an empty stomach
- taking the medication with the right precautions, for example, simvastatin not to be taken with grapefruit juice

A breakdown in any one of these factors has the potential to result in side effects and complications for the patient thus resulting in poorer than expected clinical outcomes, reduced quality of life, and deterioration of health. Most commonly the term adherence is used with respect to self-administered oral medications, but it may also apply to the use of medically prescribed devices, for example, inhalers, counseling sessions, or exercises. The term nonadherence tends to imply patients not taking any medication. In reality, there are not simply these two dichotomous extremes (adherence vs nonadherence); adherence can vary along a continuum from 0 to more than 100%, since patients sometimes take more than the prescribed amount of medication.[2] This is problematic because safety is impaired if patients are taking too much of a medication, which is cited as more of an issue for oncology patients taking oral chemotherapy drugs and who have adopted a "more is better" approach.[7,8]

IS MEDICATION ADHERENCE REALLY A PROBLEM?

The World Health Organization in 2003 reported that adherence among patients typically averages only 50% and that medication nonadherence is "a worldwide problem of striking magnitude."[1] This problem is further exemplified by the key statistics related to

Table 1.1 Key Statistics Related to Medication Adherence	
Deaths	
Unnecessary deaths in the United States due to medication nonadherence	~125,000
Unnecessary deaths in the EU due to medication nonadherence	~200,000
Costs	
Avoidable healthcare costs to US healthcare systems due to medication nonadherence	$100 billion–$300 billion/year
Avoidable healthcare costs to EU healthcare systems due to medication nonadherence	€125 billion/year
Hospital readmissions within 30 days due to medication nonadherence in the United States	~64%
Loss of revenue to US pharmaceutical companies due to unfilled prescriptions	$188 billion/year
Wastage	
Cost of unused (wastage) NHS prescription medicines in the United Kingdom	£4 billion/year
Cost of unused (wastage) prescription medicines in Canada	$8 billion/year
Prevalence	
Worldwide the number of tablet medications not taken as recommended	~50%
Adherence rates for cardiovascular therapy drugs	~50%
Adherence rates for different oral cancer therapy drugs	14–97%
Adherence rates for oral diabetes medications	36–93%
Adherence rates for antidepressant therapy	25–50%
Adherence rates for schizophrenia drugs	11–80%
Adherence rates for HIV/AIDS drugs	70–95%
Adherence rates for asthma drugs	30–70%
Adherence rates for immunosuppressant drugs	25–100%

medication adherence shown in Table 1.1. Generally the limit between good and poor adherence to prescribed drug therapy is set at 80%[9], although for individual patients the degree of nonadherence that impacts on health outcomes will vary and will be dependent on multiple factors. These include the patient's health condition, its severity, the risk of recurrence, the medication dose, and frequency of administration.

Adherence is a key factor associated with the effectiveness of all pharmacological therapies but is particularly critical for medications prescribed for chronic diseases. Furthermore, patients with multiple chronic diseases are at a significantly greater risk of medication nonadherence.[10] These problems can only increase as the world's population lives longer.

Across Europe, there is substantial regional variation between 40% and 87% adherence with prescribed pharmacotherapy, whereas in the United States the figure is approximately 50%.[11] But average figures do not give a real feel for the wide variations that occur between disease states and patient populations. For example, for patients with cancer, published results show that adherence rates for oral therapies differ widely depending on the illness; from low 14−27% for hematological malignancies to 53−98% for breast cancer and 97% for ovarian cancer.[8,12−14]

There is evidence that average consumption rates for prescribed cardiovascular drugs are reportedly only approximately 50%.[15−17] These low rates of adherence to prescribed cardiovascular therapies are likely to contribute to poor blood pressure control and lead to poor clinical outcomes. In the United States after hospitalization for acute myocardial infarction, less than 75% of patients collect their medication within 7 days of discharge. Furthermore, 34% of patients stopped taking medication within a month of being discharged from hospital. Consistent use of medication over a 6- to 12-month period was found to be low: 44% adherent for statins and 46% adherent to β-blockers.[17,18] In India, where the socioeconomic conditions are different from those in developed countries, nonadherence to hypertension drugs remains a major factor with differences in access to medications in cities compared to rural areas having an impact.[19] A study carried out in China to assess adherence to antihypertensive drugs reported a 65% level of medication adherence among hypertensive patients.[20]

Many patients who have type 2 diabetes mellitus require several different medications and adherence rates in such patients with multiple prescriptions range from 36−93% for oral hypoglycemic agents and 62−64% for insulin.[21,22]

Depression is a relatively common clinical disorder and can be difficult to treat effectively. Patient adherence with antidepressant therapy is therefore a critical aspect of effective clinical management and adherence rates of only 25−50% have been reported.[23] Furthermore, differences in medication adherence rates are documented between psychiatric populations (52%) and primary care populations (46.2%) in addition to 50% of patients discontinuing antidepressant therapy prematurely.[24]

Adherence to asthma medication regimens tends to be very poor, with the reported rates of adherence ranging from 30−70%.[25,26] Up to three-quarters of the total costs associated with asthma may be due to poor asthma control.

For human immunodeficiency virus (HIV) patients, adherence to anti-retroviral therapy is crucial to treatment outcomes because it affects how well these HIV medications decrease the patient's viral load. The lower the patient's viral load, the healthier they are likely to be. However the complexity of regimens makes adherence to therapy difficult. Adherence to HIV medications also helps to prevent drug resistance. If a patient skips a dose of their HIV medication—even once—the virus can take that opportunity to replicate and the level increases in the blood. Average adherence rates of 70–95% have been reported in Veterans Affairs Medical Centres in the United States, but adherence rates consistently greater than 95% are necessary to achieve optimal viral suppression.[27]

The range of chronic diseases cited here shows that medication non-adherence is common across diseases associated with a wide range of socioeconomic groups and not just restricted to the poorer members of society, and this is expanded upon in Chapter 4.

CONSEQUENCES OF MEDICATION NONADHERENCE

The inability of patients to take their medications in the right way has huge impact on the patients themselves, those around them and society at large.[28] Medication nonadherence is an important public health issue, affecting health outcomes and overall healthcare costs.[29] It is a growing concern to clinicians, healthcare systems, and other stake-holders, such as the pharmaceutical industry, due to its high prevalence and given the increasing occurrence of chronic diseases which require long-term pharmacotherapy.

Consequences for Patients

Suboptimal adherence to medications can have multiple consequences and a negative impact on the efficacy of treatments and patient's well-being. Nonadherence can severely impede the efficacy of oral regi-mens.[30] If a doctor is not aware that a patient is not taking a medicine as prescribed, he or she may attribute progression of the clinical condi-tion to a lack of activity of the prescribed drug, and therefore may unnecessarily change a regimen.[9] Medication nonadherence therefore limits effective management and control of chronic illnesses, increases the likelihood of preventable disease progression and treatment failure, thereby resulting in complications for the patient. In the United States, at least 125,000 people die annually due to nonadherence to medications,[31] and in the EU, it is contributing to the premature deaths of nearly 200,000 Europeans annually.[32,33]

Worldwide the elderly population are a cohort in which the prevalence of multiple chronic illnesses (multimorbidities) is high and continues to increase.[34] Polypharmacy, defined as the use of four or more regular medications by one individual, is most common in the elderly who are suffering from multiple chronic conditions.[35] These elderly patients would benefit the most from taking medications correctly but due to the complexity of having to take different tablets at different times they are the most at risk from failing to take them properly. Among older adults the consequences of medication nonadherence may be more serious, less easily detected, and less easily resolved than in younger groups.[36,37] Medication nonadherence is also common in children and adolescents with chronic illnesses.[38]

Consequences for Healthcare Systems
The costs of medication nonadherence are staggering and growing. This is because nonadherence with medication regimes results in additional use of scarce healthcare resources, such as avoidable doctor visits, laboratory tests, unnecessary additional treatments, and hospital or nursing home admissions all contributing to significantly increased healthcare costs.

For instance, poor medication adherence results in 33–69% of medication-related hospital admissions in the United States, at a cost of between $100 billion and $300 billion per year, representing 3–10% of total US healthcare costs.[2,39,40] An estimated 10% of hospitalizations in older adults may be caused by medication nonadherence.[41] More recently, a report published by the New England Healthcare Institute (NEHI) estimated that $290 billion in avoidable medical spending across all chronic diseases could be attributed to medicine-related problems, of which nonadherence to medications was a subset.[42] Examples of the benefits of the better use of medicines have been highlighted in the US PhRMA report 2015. It was estimated that suboptimal prescribing and medication errors led to additional avoidable costs to the health service of around $213 billion. Patients who were nonadherent to antihypertensives over a 3-year period were 7%

(year 1), 13% (year 2), and 42% (year 3), respectively, more likely to develop coronary artery disease and chronic heart failure than adherent patients. Reduction in cholesterol levels by statin therapy is associated with 40,000 fewer deaths, 60,000 fewer hospitalizations for heart attacks, and 22,000 fewer for strokes in 1 year with a gross saving of $5 billion. Better access to diabetes medication and improved adherence has led to 1 million fewer hospital emergency department visits and a saving of $8.3 billion. Patients with multiple conditions including, for example, hypertension, high cholesterol, diabetes, or COPD who adhered to their prescription regimens had fewer days off work than their nonadherent colleagues.[41]

Medication nonadherence is costing EU governments an estimated €125 billion annually.[34] Five percent of all emergency admissions in the United Kingdom are for people who do not take their medicines as prescribed, and this is estimated to cost the National Health Service (NHS) in England £500 million annually.[43,44] These estimates do not even capture the costs associated with time off from work, and other types of expenses incurred by family and support networks.

In the United States approximately 64% of hospital readmissions within 30 days of discharge are attributed to medication adherence issues.[45] Improving medication adherence is therefore one facet of the Medicare Hospitals Re-admissions Reduction Program, which provides financial incentives to hospitals to lower readmission rates occurring within 30 days of discharge.[46] Currently the financial incentives (penalties) may amount to a 2% reduction in the total patient charges paid to the hospital based on the readmission of Medicare patients who originally went into hospital with one of three conditions: heart attack, heart failure, or pneumonia. In the United Kingdom there is also concern over the 30-day readmission rates and a similar use of financial incentives to improve the situation has been introduced.[47] The NHS has opted to specify which disease states will be excluded from this protocol and these include cancer care, all children under 17, maternity, mental health, end of life care, and definitive treatment adjustment.

Such concerns are not restricted to the United States and United Kingdom. In Ireland, for example, the Irish Platform for Patients' Organisations, Science and Industry (IPPOSI) has also identified medication nonadherence as an important issue to highlight, discuss, and tackle.[48] The Western Australian Medication Adherence and Costs in

Heart (WAMACH) disease study has evaluated the clinical and economic consequences of nonadherence to secondary prevention drugs for cardiac disease in 40,000 coronary heart disease patients aged 65 years or older.[49] The conclusion of this study indicated that there was a need for long-term adherence monitoring of medications for cardiac patients and this information would be of interest to policymakers, patients, and healthcare providers.

Medicines Wastage

Medicines wastage is another consequence of medication nonadherence and can be categorized as either therapeutic loss and/or material waste.[50,51] Therapeutic loss occurs where the effects of the medicine(s) are reduced or negated by the user's failure to take them as prescribed. Material waste occurs where the medicines are physically unused and either disposed of, returned to the pharmacy, or stock piled in the patient's home. In the United Kingdom the National Institute for Clinical Excellence (NICE) produced guidelines for medication adherence in which it estimated that approximately £4 billion of medicines supplied on prescription through the NHS are not used correctly.[52] It is estimated that 3–7% of medications intended for patients are unused in the United States and that similar wastage could cost Canada approximately $8 billion annually.[53] In several African countries the magnitude of medicines wastage is less known since in these countries medicines which remain unused due to various reasons such as lack of adherence are normally retained in the community for future use.[54]

The wastage of medicines due to poor adherence also impacts on the pharmaceutical industry since it takes many years for them to develop a new medicine to the standards of quality, efficacy, and safety laid down by legislation and to do all of the research and development necessary before a medicine can be licensed for use. It costs billions to develop a new medicine and this is wasted if patients do not take their medications correctly and therefore do not gain maximum benefit from the prescribed drug therapy (Fig. 1.1). For example in 2012 US pharmaceutical companies lost an estimated $188 billion annually in revenue because patients failed to refill their medications as directed.[55] Indeed the World Health Organization has declared that more people worldwide would benefit from efforts to improve medication adherence rather than the development of new medical treatments.[1] This will avoid waste of the

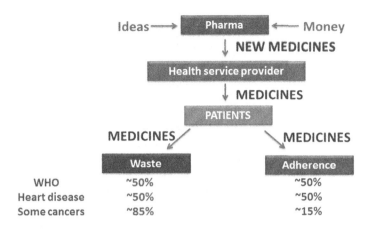

Figure 1.1 Medicines wastage associated with medication adherence.

pharmaceutical industries' efforts, the health service provider's resources and more importantly will improve overall patient well-being. At a point when care is ever more dependent on medication therapy for treating an evergrowing number of patients improving medication adherence is possibly the major challenge for healthcare systems. The responsibility for improving adherence has usually been placed on the patient and healthcare provider. However, if the problem of medication adherence is to be more effectively addressed, other members of the healthcare and pharmaceutical industries must also take responsibility.

Understanding the magnitude and scope of the problem of medication adherence is the first step in reaching improved adherence rates. The second step is to evaluate the risk factors for each patient that affect medication adherence/nonadherence.[56,57]

FACTORS AFFECTING MEDICATION ADHERENCE

Medication nonadherence can be unintentional, for example, forgetting to take a dose of a medicine. It can sometimes be intentional, for example, deliberately skipping a dose to try to avoid side effects or because of concerns about the expense of the drug. It can be defined by several behavioral patterns, including failure to collect prescriptions, failure to follow day-to-day instructions (eg, taking too few or too many doses, or taking medication with inappropriate food) and failure to collect subsequent prescriptions as directed. Either way, the outcomes for the patient can be risky.[2,3,6]

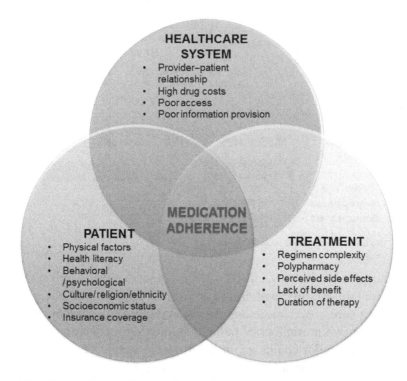

Figure 1.2 A subdivision of factors affecting medication adherence.

Medication nonadherence is a multifactorial problem that can be influenced by the interrelationship of various factors. According to the WHO, these factors are in the following categories: social and economic, healthcare system, health condition-related, therapy-related, and patient-related.[1] In broader terms, these factors fall into the categories of patient factors, treatment factors, and healthcare system factors, thus impinging on social and administrative aspects of pharmacy and medicine (Fig. 1.2).

Patient Factors

Several patient-related factors are determinants of medication adherence.[3,15,58,59] These factors can be further divided into demographic, sociocultural, and behavioral factors[29] and include:

- forgetting to take medicine
- cultural, religious, and lay beliefs about illness
- denial of illness
- lack of understanding of disease
- mental health issues

- physical factors including visual impairment, hearing impairment, impaired mobility, swallowing problems
- cognitive function
- lack of involvement in the treatment decision-making process
- demographic factors including age, sex, education, employment, income, marital status, and ethnic status
- limited language proficiency especially with respect to health concerns
- poor health literacy
- poor social support network
- the assumption that medication can be discontinued if one feels better
- medicine not covered by insurance or lack of healthcare insurance
- motivation
- stress, anxiety, and anger
- alcohol or substance abuse
- selling on prescription medicines
- "doctor shopping"

Treatment Factors[16,29,59]

- Regimen complexity including number of daily doses and poly-pharmacy
- actual or perceived side effects of medicine
- high drug costs
- duration of therapy
- frequent changes in medication regimen
- perception that a prescribed medication would have little benefit
- medications with social stigma attached to use
- treatment interferes with lifestyle or requires significant behavioral changes
- medication storage requirements
- lack of symptoms with a chronic illness
- severity of symptoms
- counterfeit medicines, for example, malaria treatment tablets and HIV/AIDS treatment tablets in Sub-Saharan Africa

Healthcare System Factors[16,29,59]

- Poor quality of provider–patient relationship
- poor provider communication skills (contributing to lack of patient knowledge or understanding of the treatment regimen)

- prohibitive drug costs, copayments, or both
- lack of positive reinforcement from the healthcare provider
- weak capacity of the system to educate patients and provide follow-up
- patient information leaflets written at too high literacy level
- limit access to care (making appointments difficult to schedule)
- long wait times
- lack of continuity of care

There are many interrelationships that exist among patient, treatment, and healthcare system–related factors and there is also clear evidence that medication nonadherence is a complex problem.[60] A recent review by Mathes et al.[61] concluded that only a few factors had a consistent influence on adherence. Furthermore, belonging to an ethnic minority, unemployment and cost for the patient for their medications showed a negative effect on medication adherence, which indicate further that social aspects are involved. Given such degrees of complexity, it is not surprising that some practice guidelines to improve adherence have been issued globally.

CLINICAL PRACTICE GUIDELINES FOR THE IMPROVEMENT OF MEDICATION ADHERENCE

Improving medication adherence, potentially the most effective route to improving the therapeutic benefit of pharmacotherapy, remains a challenge for healthcare systems worldwide. A recent review evaluated national and international clinical practice guidelines designed to help healthcare providers address patients' medication adherence and identified 23 guidelines of varying detail and quality.[62] Nine of the 23 guidelines originated from the United States, three from Canada, three from the United Kingdom, and one each from Australia and Spain. Six guidelines had authors from multiple countries in specific regions—one from Central and South America, one in the Middle East, and one in Europe. The guidelines' treatment foci were:

- 5 guidelines for HIV/AIDS
- 3 guidelines for hypertension
- 2 guidelines for cardiovascular disease
- 2 guidelines for contraception

- 2 guidelines for menopause
- 1 guideline for osteoporosis
- 1 guideline for renal transplant
- 1 guideline for mental health
- 1 guideline for depression
- 1 guideline for asthma
- 1 guideline for multiple sclerosis
- 3 guidelines did not specify a health condition

Surprisingly, there were no guidelines for assessing adherence to pain management drugs which is a major health concern.[63] Furthermore, none of the published guidelines reported the use of any direct assessment methodology for monitoring adherence to prescribed drug therapy and in their "What is Missing" section the authors failed to highlight the opportunity that direct measurements could offer.[62]

Clinicians need measures to assess adherence to prescribed pharmacotherapy to aid the clinical decision-making process in the event of poor patient progress and to maximize the patient health outcomes from the drug therapies prescribed.

METHODS FOR ASSESSING MEDICATION ADHERENCE

A multitude of methods have been used to assess medication adherence in patients although no "gold standard" exists for use in routine clinical practice.[58] For the adherence measures identified, each have their advantages and limitations as summarized in Table 1.2.[2,56,61,64−67]

Indirect Assessment Methods

Adherence to medication can be measured by indirect assessment methods which include pill counts, patient questionnaires, electronic monitors, patient self-reports, and prescription refill rates. While these assessments are relatively easy to carry out and are the most commonly used,[61] they do not always provide the required information and are proxy measures of medication adherence. For instance, pill counts do not provide information of other aspects of taking medications, such as dose timing which may be important in determining clinical outcomes. More recently, to eliminate the tendency toward

Table 1.2 Indirect and Direct Methods Used to Assess Medication Adherence[2,56,61,64–67]

Method	Comments	Outcomes
Indirect Methods		
Patient interview	Noninvasive/easy/inexpensive/global	Optimistic/interviewer dependent/too subjective
Patient diary	Noninvasive/self-report/inexpensive/global	Optimistic/no confirmation of use/diary must be returned
Pill count	Noninvasive/easy/inexpensive/global	No confirmation of use or adherence/pills may be lost or sold
Questionnaire	Noninvasive/must collect data/global	No confirmation of use/no continuous data
Electronic monitoring	Noninvasive/expensive/globally difficult	May be manipulated/does not suit all pharmaceutical dosage forms
Prescription refills	Noninvasive/long-term records/globally difficult	Medication collection confirmed/no confirmation of ingestion
Clinical outcomes	Affected by other factors	Adherence is a surrogate endpoint of clinical outcomes
Direct Methods		
Directly observed therapy	Noninvasive/time consuming/global/medication can be hidden in mouth/impractical in outpatient setting	Confirmation that medication has been taken only during clinics
Blood medication/metabolite monitoring	Invasive/costly transport & technique/globally difficult	Confirms recent use/patient PK data/measured blood drug levels/objective
Dried blood spot analysis medication/metabolite monitoring	Minimally invasive/easy transport/expensive equipment/global	Confirms recent use/patient PK data/measured blood drug levels/objective
Urine medication/metabolite monitoring	Noninvasive/costly transport & analysis/globally difficult	Confirms recent use/patient PK data/objective
Biomarkers in blood	Invasive/expensive equipment/globally difficult/influenced by other biological parameters, drugs and diet	Confirms recent use/patient PK data/derived blood drug levels/biomarker not available for all drugs
Ingestible medication marker	Limited research on its use/costly/not designed for clinical practice	Ingestible tiny mircosensor fixed in each tablet

optimistic patient-based reports, a pilot study in Spain was conducted which required patients to send a short video of them taking their medication to their healthcare provider using a mobile phone.[68] Other recently investigated approaches are technology-derived and include:

- mobile phone real-time assessment, monitoring, and alerts[69–72]
- smart pills with ingestible sensors and packaging that flashes at medication time[73]
- web-based games to help manage the condition and side effects[74]
- computerized logbooks[69]

The efficacy of primary healthcare clinic-based 'medication adherence clubs' to help patients suffering from multiple chronic diseases in a resource-limited setting has also been investigated.[75]

In the United States several pharmaceutical companies have taken the initiative to introduce patient clubs or groups to monitor and improve adherence to specific medicines. These appear to follow a general pattern of the provision of free advice by phone or online, in conjunction with discounted medicine costs for continuous use of the companies' products or based on improvements to the medical condition. Benefits for the patients, from these activities, included adherence to particular diabetes medicines rising to 87%, better control of blood pressure and, for the companies, an increase in brand loyalty, and in one instance a rise in sales of 510%.[73] Despite these successes, this approach is limited to countries where patients pay directly for the medicines. In the United Kingdom, for example, one of the main medicine user groups, elderly people, are exempt from medicine charges and the problems with nonadherence need alternative approaches.

Direct Assessment Methods

Direct assessment methods include direct patient observation, determination of the level of a drug or its metabolite in blood or urine, measurement of biomarker in blood or urine, and the detection of an ingestible medication marker, added to the dosage form, in the blood. Such direct approaches are some of the most accurate methods of measuring medication adherence; however, current measures are more costly in terms of both patient and clinician time and acquiring such biological samples requires a visit to a clinic or hospital. The costs associated with direct assessment can be reduced without detriment to the information produced, by the use of a finger prick blood sample collected as a dried blood spot (DBS) for the determination of drug or biomarker levels as a measure of medication adherence.[76] Moreover, direct assessment based on therapeutic drug, metabolite, or biomarker monitoring of plasma, serum, DBS, or urine samples also provide more objective measures; however, such levels may vary because of variations in patient pharmacokinetics and pharmacogenetics. Furthermore the impact of urine sample collection time on results has been reported as having "white coat compliance" in which improvements in medication adherence several days prior to a scheduled medical examination was observed.[66] Due to the many advantages offered by saliva and hair sampling compared with blood and urine sampling, these biosamples have been investigated in medication adherence studies.[77,78]

Tanna and Lawson have suggested that monitoring blood drug levels was the more appropriate way forward for ensuring drug(s) are present within the therapeutic window for individual patients.[76] This concept is reenforced by Morrison et al.[79] who suggested that the level of medication adherence is defined by the patient drug levels in blood remaining within the therapeutic window irrespective of when the drug dose was taken. In reality, this approach can only be put into practice if direct assessment methods are utilized.

Given the high prevalence and detrimental impact of medication nonadherence on clinical and economic outcomes, there is a clear need for guidance on the direct measurement of medicines in patient biosamples. Healthcare professionals such as clinicians, pharmacists, and nurses have major roles in their daily practice to improve adherence to medications for patients. A basic need for addressing the global problem of medication adherence is the accurate measurement of drugs in biological samples to provide healthcare professionals with a reliable base for decision-making. Only by the direct analysis of a patient biosample can this confidence be achieved.

CONCLUSIONS

Nonadherence to prescribed drug therapy is a formidable and widespread problem often leading to a reduction in or lack of treatment benefits, extra visits to the doctor, and unnecessary hospital admissions. This chapter has highlighted the critical waste of medicines and the extra unnecessary costs borne by health service providers as a result of patient nonadherence to prescribed medicines. A recent investigation[80] of the implementation of medication adherence policy solutions in 10 EU countries found that the policy leaders reported that there was insufficient implementation of the initiatives at patient, government, or healthcare provider levels. This means that patients still do not receive the optimum benefit from the prescribed medication despite many attempts to involve them more in following the care regimen by indirect assessment methods.

The use of direct assessment of patients' adherence to prescribed regimens is viewed differently around the world: from laboratories advertising such assessments for a few dollars in the United States to initial trials of "at home tests" in the EU, to "you can't do that—it's Big Brother" in the United Kingdom. With this diverse range of attitudes perhaps it is better to seek the patient's feelings on providing additional biosamples. The patients' response at a focus group meeting held at De Montfort University in the United Kingdom was "if this makes us better quicker—we are all for it!" This patient acceptance of the approach using analytical chemistry to assess medication adherence may help to stimulate the stalled progress seen in the EU.[80]

REFERENCES

1. Sabaté E. *Adherence to Long-Term Therapies: Evidence for Action*. Geneva, Switzerland: World Health Organization; 2003.

2. Osterberg L, Blaschke T. Adherence to medication. *N Engl J Med*. 2005;353(5):487–497.

3. Brown MT, Bussell JK. Medication adherence: WHO cares? *Mayo Clin Proc*. 2011;86 (4):304–314. Available from: http://dx.doi.org/10.4065/mcp.2010.0575.

4. Horne R. Compliance, adherence and concordance: implication for asthma treatment. *Chest*. 2006;130:65–72.

5. Hansen RA, Farley JF, Droege M, Maciejewski ML. A retrospective cohort study of economic outcomes and adherence to monotherapy with metformin, pioglitazone, or a sulfonylurea among patients with type 2 diabetes mellitus in the United States from 2003 to 2005. *Clin Ther*. 2010;32(7):1308–1319.

6. *Patient Resource*. Medication adherence. <http://www.patientresource.com/medication_adherence_overview.aspx>; 2015 Accessed 22.09.15.

7. Weingart SN, Flug J, Brouillard D, et al. Oral chemotherapy safety practices at US cancer centres: questionnaire survey. *BMJ*. 2007;34:407.

8. Partridge AH, Avorn J, Wang PS, Winer EP. Adherence to therapy with oral antineoplastic agents. *J Natl Cancer Inst*. 2002;94(9):652–661.

9. Avorn J, Monette J, Lacour A, et al. Persistence of use of lipid-lowering medications: a cross-national study. *JAMA*. 1998;279(18):1458–1462.

10. Vogeli C, Shields AE, Lee TA, et al. Multiple chronic conditions: prevalence, health consequences, and implications for quality, care management, and costs. *J Gen Intern Med*. 2007;22(suppl 3):391–395.

11. Larsen J, Stovring H, Kragstrup J, Hansen DG. Can differences in medical drug compliance between European countries be explained by social factors: analyses based on data from the European Social Survey, round 2. *BMC Public Health*. 2009;9:145. Available from: http://dx.doi.org/10.1186/1471-2458-9-145.

12. D'Amato S. Improving patient adherence with oral chemotherapy. *Oncol Issue*. 2008;42–45: July/August issue.

13. Ruddy K, Mayer E, Partridge A. Patient adherence and persistence with oral anticancer treatment. *CA Cancer J Clin*. 2009;59(1):56–66. Available from: http://dx.doi.org/10.3322/caac.20004.

14. Mathes T, Antoine SL, Pieper D, Eikermann M. Adherence enhancing interventions for oral anticancer agents: a systematic review. *Cancer Treat Rev*. 2014;40(1):102–108. Available from: http://dx.doi.org/10.1016/j.ctrv.2013.07.004.

15. Baroletti S, Dell'Orfano H. Medication adherence in cardiovascular disease. *Circulation.* 2010;121:1455–1458.

16. Ho PM, Bryson CL, Rumsfeld JS. Medication adherence: its importance in cardiovascular outcomes. *Circulation.* 2009;119(23):3028–3035. Available from: http://dx.doi.org/10.1161/CIRCULATIONAHA.108.768986.

17. Kronish IM, Ye S. Adherence to cardiovascular medications: lessons learned and future directions. *Prog Cardiovasc Dis.* 2013;55(6):590–600.

18. Jackevicius CA, Li P, Tu JV. Prevalence, predictors, and outcomes of primary nonadherence after acute myocardial infarction. *Circulation.* 2008;117:1028–1036.

19. Thomas D, Meera NK, Binny K, et al. Medication adherence and associated barriers in hypertension management in India. *CVD Prev Control.* 2011;6:9–13.

20. Lee GKY, Wang HHX, Liu KQL, et al. Determinants of medication adherence to antihypertensive medications among a Chinese population using Morisky Medication Adherence Scale. *PLoS One.* 2013;8(4):e62775. Available from: http://dx.doi.org/10.1371/journal.pone.0062775.

21. Cramer J. A systemic review of adherence with medications for diabetes. *Diabetes Care.* 2004;27(5):1218–1224.

22. Bailey CJ, Kodack M. Patient adherence to medication requirements for therapy of type 2 diabetes. *Int J Clin Pract.* 2011;65(3):314–322. Available from: http://dx.doi.org/10.1111/j.1742-1241.2010.02544.x.

23. Trivedi MH, Lin EH, Katon WJ. Consensus recommendations for improving adherence, self-management, and outcomes in patients with depression. *CNS Spectr.* 2007;12:S1–27.

24. Sansone RA, Sansone LA. Antidepressant adherence: are patients taking their medications? *Innov Clin Neurosci.* 2012;9(5–6):41–46.

25. Bender BG, Bender SE. Patient-identified barriers to asthma treatment adherence: responses to interviews, focus groups, and questionnaires. *Immunol Allergy Clin North Am.* 2005;25(1):107.

26. Rand CS, Wise RA. Measuring adherence to asthma medication regimens. *Am J Respir Crit Care Med.* 1994;149:S69–S76.

27. Paterson DL, Swindells S, Mohr J, et al. Adherence to protease inhibitor therapy and outcomes in patients with HIV infection. *Ann Intern Med.* 2000;133:21–30.

28. Stilley CS, Bender CM, Dunbar-Jacob J, Sereika S, Ryan CM. The impact of cognitive function on medication management: three studies. *Health Psychol.* 2010;29(1):50–55. Available from: http://dx.doi.org/10.1037/a0016940.

29. Iuga AO, McGuire MJ. Adherence and health care costs. *Risk Manag Healthc Policy.* 2014;7:35–44.

30. Bedell CH. A changing paradigm for cancer treatment: the advent of oral chemotherapy agents. *Clin J Oncol Nurs.* 2003;7:5–9.

31. McCarthy R. The price you pay for the drug not taken. *Bus Health.* 1998;16:27–33.

32. PGEU GPEU Statement—Targeting adherence. Brussels; 2008.

33. European Council Policy Makers Debate. An EU response to medication non-adherence. Brussels; 2010.

34. Koné Pefoyo AJ, Bronskill SE, Gruneir A, et al. The increasing burden and complexity of multimorbidity. *BMC Public Health.* 2015;15:415. Available from: http://dx.doi.org/10.1186/s12889-015-1733-2.

35. Maher RL, Hanlon JT, Hajjar ER. Clinical consequences of polypharmacy in elderly. *Expert Opin Drug Saf.* 2014;13(1). Available from: http://dx.doi.org/10.1517/14740338.2013.827660.

36. Hammarlund ER, Ostrum JR, Kethley AJ. The effects of drug counseling and other educational strategies on drug utilization of the elderly. *Med Care*. 1985;23(2):165–170.

37. Banning M. Older people and adherence with medication: a review of the literature. *Int J Nurs Stud*. 2008;45(10):1550–1561.

38. Dean AJ, Walters J, Hall A. A systematic review of interventions to enhance medication adherence in children and adolescents with chronic illness. *Arch Dis Child*. 2010;95:717–723. Available from: http://dx.doi.org/10.1136/adc.2009.175125.

39. *IMS Institute for Healthcare Informatics*. Avoidable costs in US health care. <https://www.imshealth.com/deployedfiles/imshealth/Global/Content/Corporate/IMS%20Institute/RUOM-2013/IHII_Responsible_Use_Medicines_2013.pdf>; 2013 Accessed 12.10.15.

40. Benjamin RM. Medication adherence: helping patients take their medicines as directed. *Public Health Rep*. 2012;127(1):2–3.

41. *Pharmaceutical Research and Manufacturers of America*. 2015 biopharmaceutical research industry profile. <http://www.phrma.org/profiles-reports>; April 2015 Accessed 28.10.15.

42. *New England Health Institute (NEHI)*. Thinking outside the pillbox: a system-wide approach to improving patient medication adherence for chronic disease; 2009. Cambridge, MA.

43. Up in smoke: £150m of medicines a year. <http://www.theguardian.com/society/2012/jul/03/prescription-medicine-unused>; 2012 Accessed 08.10.15.

44. Barnett NL. Medication adherence: where are we now? A UK perspective. *Eur J Hosp Pharm*. 2014;21:181–184.

45. *Healthcare Intelligence Network*. 2013 Healthcare benchmarks: improving medication adherence. <http://www.hin.com/library/2013HealthcareBenchmarksMedicationAdherence_preview.pdf>; 2013 Accessed 28.01.16.

46. Rau J. Medicare to cut pay to 2,225 hospitals over readmissions. <http://www.medpageto-day.com/HospitalBasedMedicine/GeneralHospitalPractice/40821>; 2013 Accessed 08.10.15.

47. *NHS Confederation briefing*. The impact of non-payment for acute readmissions. <http://www.chks.co.uk/userfiles/files/The%20impact%20of%20non-payment%20for%20acute%20readmissions%20FINAL%20FOR%20WEB.pdf>; 2011;Issue 211 Accessed 08.10.15.

48. *Irish Platform for Patients' Organisations, Science and Industry*. A focus on patient compliance and adherence in 2013. *Outcome report*. <http://www.ipposi.ie/index.php/news-a-events-mainmenu-28/247-follow-on-from-rare-disorder-without-border>; 2013 Accessed 08.10.15.

49. Gunnell AS, Knuiman MW, Geelhoed E, et al. Long-term use and cost-effectiveness of secondary prevention drugs for heart disease in Western Australian seniors (WAMACH): a study protocol. *BMJ Open*. 2014;4(9):e006258. Available from: http://dx.doi.org/10.1136/bmjopen-2014-006258.

50. Hazell B, Robson R. Pharmaceutical waste reduction in the NHS. *NHS England report* 2015; Version 1.

51. Bain T. Public health implications of household pharmaceutical waste in the United States. *Health Serv Insights*. 2010;3:21.

52. *The National Institute for Clinical Excellence (NICE)*. Costing statement: medicines adherence: involving patients in decisions about prescribed medicines and supporting adherence. <https://www.nice.org.uk/guidance/cg76/resources/costing-statement-242056477>; 2009 Accessed 27.10.15.

53. Tchen J, Vaillancourt R, Pouliot A. Wasted medications, wasted resource. *Can Pharm J*. 2013;146(4):181–182. Available from: http://dx.doi.org/10.1177/1715163513490654.

54. Kagashe GA, Makenya FB, Buma D. Medicines wastage at a tertiary hospital in Dar Es Salaam Tanzania. *J Appl Pharm Sci*. 2014;4(06):098–102.

55. Congdon K. Pharma needs a medication adherence game plan. *Pharmaceutical online*, <http://www.pharmaceuticalonline.com/doc/pharma-needs-a-medication-adherence-game-plan-0001>; 2015 Accessed 22.10.15.

56. Neiheisel MB, Wheeler KJ, Roberts ME. Medication adherence part one: understanding and assessing the problem. *J Am Assoc Nurse Pract*. 2014;26:49−55.

57. Wheeler KJ, Roberts ME, Neiheisel MB. Medication adherence part two: predictors of non-adherence and adherence. *J Am Assoc Nurse Pract*. 2014;26:225−232.

58. Kennedy J, Tuleu I, Mackay K. Unfilled prescriptions of medicare beneficiaries: prevalence, reasons, and types of medicines prescribed. *J Manag Care Pharm*. 2008;14(6):553−560.

59. *Adult Medication*™. Improving medication adherence in older adults. Overview of medication adherence—where are we today? <http://www.adultmeducation.com/OverviewofMedicationAdherence.html>; 2006 Accessed 14/10/15.

60. Kardas P, Lewek P, Matyjaszczyk M. Determinants of patient adherence: a review of systematic reviews. *Front Pharmacol*. 2013;4:91. Available from: http://dx.doi.org/10.3389/fphar.2013.00091.

61. Mathes T, Jaschinski T, Pieper D. Adherence influencing factors—a systemic review of systemic reviews. *Arch Public Health*. 2014;72:37.

62. Ruppar TM, Dobbels F, Lewek P, et al. Systematic review of clinical practice guidelines for the improvement of medication adherence. *Int J Behav Med*. 2015;. Available from: http://dx.doi.org/10.1007/s12529-015-9479-x.

63. Melanson SE, Ptolemy AS, Wasan AD. Optimizing urine drug testing for monitoring medication compliance in pain management. *Pain Med*. 2013;14(12):1813−1820. Available from: http://dx.doi.org/10.1111/pme.12207.

64. Lehmann A, Aslani P, Ahmed R, et al. Assessing medication adherence: options to consider. *Int J Clin Pharmacol*. 2014;36(1):55−69. Available from: http://dx.doi.org/10.1007/s11096-013-9865-x.

65. Farmer KC. Methods for measuring and monitoring medication regimen adherence in clinical trials and clinical practice. *Clin Ther*. 1999;21:1074−1090.

66. MacLaughlin EJ, Raehl CL, Treadway AK, et al. Assessing medication adherence in the elderly: which tools to use in clinical practice? *Drugs Aging*. 2005;22(3):231−255.

67. Hawkshead J, Krousel-Wood MA. Techniques for measuring medication adherence in hypertensive patients in outpatient settings. Advantages and limitations. *Dis Manage Health Outcomes*. 2007;15(2):109−118.

68. *European Lung Foundation*. Virtual monitoring could aid adherence to TB medication. <http://www.europeanlung.org/en/news-and-events/media-centre/press-releases/virtual-monitoring-could-aid-adherence-to-tb-medication>; 2013 Accessed 15.10.15.

69. Clifford S, Perez-Nieves M, Skalicky AM, Reaney M, Coyne KS. A systematic literature review of methodologies used to assess medication adherence in patients with diabetes. *Curr Med Res Opin*. 2014;30(6):1071−1085.

70. Wark PA, Car J. Review suggests mobile phone interventions improve medication adherence; rigorous longer term studies are needed to confirm effects. *Evid Based Nurs*. 2015;18:120. Available from: http://dx.doi.org/10.1136/eb-2014-102010.

71. Stoner SA, Arenelia PB, Hendershot CS. Randomized controlled trial of a mobile phone intervention for improving adherence to naltrexone for alcohol use disorders. *PLoS One*. 2015;10(4):e0124613.

72. van Boven JFM, Trappenburg JCA, van der Molen T, Chavannes NH. Towards tailored and targeted adherence assessment to optimise asthma management. *NPJ Prim Care Respir Med*. 2015;25:15046. Available from: http://dx.doi.org/10.1038/npjpcrm.2015.46.

73. *Capegemini Consulting*. Patient adherence: the next frontier in patient care. Vision & Reality, 9th Edition.   <https://www.uk.capgemini.com/resource-file-access/resource/pdf/Patient_Adherence__ The_Next_Frontier_in_Patient_Care.pdf>; 2011 Accessed 27.10.15.

74. *HealthPrize*. The leading patient engagement and medication adherence company. <https:// healthprize.com/>; 2015 Accessed 27.10.15.

75. Khabala KB, Edwards JK, Baruni B, et al. Medication Adherence Clubs: a potential solution to managing large numbers of stable patients with multiple chronic diseases in informal settlements. *Trop Med Int Health*. 2015;20(10):1265−1270.

76. Tanna S, Lawson G. Dried blood spot analysis to assess medication adherence and to inform personalization of treatment. *Bioanalysis*. 2014;6(21):2825−2838.

77. Olds PK, Kiwanuka JP, Nansera D, et al. Assessment of HIV antiretroviral therapy adherence by measuring drug concentrations in hair among children in rural Uganda. *AIDS Care*. 2015;27(3):327−332. Available from: http://dx.doi.org/10.1080/09540121.2014.983452.

78. Moore C. Drug testing and adherence monitoring in pain management: oral fluid testing. *J Opiod Manag*. 2015;11(1):69−75. Available from: http://dx.doi.org/10.5055/jom.2015.0254.

79. Morrison A, Stauffer ME, Kaufman AS. Defining medication adherence in individual patients. *Patient Prefer. Adher*. 2015;9:893−897.

80. Clyne W, McLachlan S. A mixed-methods study of the implementation of medication adherence policy solutions: how do European countries compare? *Patient Prefer Adher*. 2015;9:1505−1515.

Opportunities and Challenges for Analytical Chemistry in the Assessment of Medication Adherence

INTRODUCTION

Medicines are one of the contributors to the increasing costs of health-care worldwide. A major challenge to the healthcare providers is to maintain these costs within affordable limits for both the patients and the providers. Inappropriate use of therapeutic drugs is a waste of the original purchase costs and may lead to additional unnecessary charges to the healthcare provider. Prescription medicines not collected may also mean a loss of revenue for the manufacturer. An accurate knowledge of adherence to prescription is therefore important to the pharmaceutical industry, the healthcare supplier, and ultimately the patient. This information can only be obtained objectively by direct analyses of biosamples obtained from patients. It is therefore in every-one's interest to make this happen.

OPPORTUNITIES

The provision of modern healthcare is a complex and expensive under-taking and yet scant attention appears to be paid ensuring that resources developed at multibillion dollar expense are used on an opti-mal and cost-effective basis in order to maximize benefit to both the patient and the healthcare system.[1] The resource referred to is of course the use of therapeutic drugs in patient care. In principle instru-mental analytical chemical measurements of drug levels in a patient biosample can be used to:

- Tackle arguably the biggest threat facing modern healthcare provision, namely, the lack of adherence to medication. As we discussed in Chapter 1 this is a worldwide problem with adherence levels typically between 30% and 70%.[1] The absence of

Analytical Chemistry for Assessing Medication Adherence. DOI: http://dx.doi.org/10.1016/B978-0-12-805463-5.00002-X

the prescribed drug in the analysis of a patient's blood sample would be a cause for concern for the clinician.

- Identify and limit the incorrect use of medicines and thereby reduce not only the direct costs of wastage involved but also the much larger and possibly unnecessary costs of re-admission to hospital.

- Enable therapeutic drug monitoring (TDM) in the optimal use of critical dose drugs. Such drugs are defined by a narrow therapeutic window in which elevated blood concentrations can cause significant toxicity, or under-dosing can result in serious consequences from ineffective treatment.[2] A critical example is the lifelong necessity to maintain the correct levels of immunosuppressant drugs in organ replacement procedures and subsequent recovery in order to prevent rejection and possible patient death.[3,4]

- Provide indications of the cause of suboptimal medication as a result of either the misdirection of prescribed drugs, adverse drug–drug interactions or the use of other nonprescribed pain control therapies. The area of pain management therapy provides good examples of these with prescription tablets being diverted to other users[5] and the consumption of additional pain killers, for example, either a therapeutic or an illegal drug.[6]

- Facilitate the personalization of treatment so that medication can be optimized based on pharmacogenetic analysis of patients.[7] It is now recognized that in some instances drug–patient interactions differ between individuals and the therapeutic effects may be more or less pronounced. Commercial and hospital based laboratories in the United States and Canada offer such tests in order to identify the drugs better suited to a particular patient based on the results obtained.

- Confirm directly to the patient that they are adhering to a pre-scribed therapy. Routine self-monitoring of blood glucose levels[8] at least twice a day is a recommended procedure for a diabetic patient. This ensures that the blood glucose level is maintained within a pre-defined range and is, in fact, an indirect measure of adherence to prescribed pharmacotherapy. In Holland, self-monitoring of lithium levels in blood[9] is a direct measure of adherence to treatment for mental disorders and is used to ensure that the level remains within the very narrow therapeutic window.

- Identify counterfeit or substandard medicines which could potentially lead to mass instances of "unknowing" poor adherence. Such a situation occurred in Pakistan where more than 100 patients died after receiving substandard tablets for the treatment of cardiovascular disease.[10]

It is apparent from the above examples that in some instances measured values are being used to assess adherence to medication, principally where failure to do so has dire consequences such as organ rejection or death. These examples are usually from intensive care situations where the collection of whole blood samples is accepted as the norm. To utilize the idea of chemical analysis as a means of assessing adherence on a much wider basis it is necessary to gain patient acceptance of a proposed scheme which must satisfy the following three challenges:

- The collection, from the patient, of a suitable biosample for analysis. Not only must the sampling strategy be acceptable to the patient but it must also provide enough of the drug to be detectable at the therapeutic concentrations anticipated. Furthermore, the proposed samples must remain stable for the duration of the transit period from patient to analytical laboratory.
- To develop an analytical capability with sufficient sensitivity and specificity to uniquely identify (screening) and quantify (assess) target drug levels in the biosample available. Such instruments must be reliable, reproducible, and able to handle many hundreds of samples per shift.
- To promote the wider understanding of the value of the information obtained especially among clinicians who will treat the patients and among the regulating authorities. There are reports showing that clinicians are not always fully aware of new developments, particularly in the assessment of adherence to medication.[11,12] Regulatory assessment of a new methodology is frequently arranged by the instrument manufacturer who anticipates recovering the costs involved from future sales.

CHALLENGES

These challenges can be redefined by stating that it is necessary to know where the drug is in the body, as a function of time after dosing, in order to collect an appropriate sample. It is also necessary to know the target analyte, the active ingredient or its metabolite, and the anticipated concentration in order to match sample size with the analytical capability available (see Chapter 3).

Understanding Pharmacokinetic Effects on the Concentration of a Drug in the Bloodstream

Understanding the basic principles of drug pharmacokinetics is crucial when identifying meaningful samples to analyze in order to evaluate adherence to a prescribed drug regimen.

ADME is an abbreviation in pharmacokinetics for absorption (A), distribution (D), metabolism (M), and excretion (E) used to describe the disposition of a pharmaceutical compound in the body. Pharmacokinetics is concerned with the variation in drug concentration with time as a result of ADME. It may be simply defined as what the body does to the drug. Readers should note that here *distribution* means where the drug is deposited in the body and not the act of moving the drug around the body.

Absorption

Absorption of drugs from the administration site into the bloodstream is the first major barrier to consider in determining the amount of drug that enters the systemic circulation. For instance, orally administered drugs must pass through the intestine and liver before they reach the bloodstream. As a result, orally administered drugs typically exhibit more pharmacokinetic (time to effect) variability than intravenously administered drugs. Absorption of a drug will vary between patients and identical doses may not therefore necessarily produce the same concentrations at the target cells. Poor solubility of the active pharmaceutical ingredient, poor chemical stability in the stomach, and an inability to penetrate the intestinal wall all combine to reduce the amount of the drug, from the original dose, which is available for therapeutic action. This fraction is known as the bioavailability of the compound. Direct inhalation is a very common route of administration with absorption directly into the bloodstream which leads to a very rapid biological effect. Other routes of administration, for example, direct injection of solutions must be considered for active ingredients with poor absorption. Other factors which affect absorption into the bloodstream are dose, diet, gastric emptying time, intestinal transit time, drug–drug interactions, and pharmacogenetic effects. Simvastatin provides a good example of the effects of both diet and drug–drug interactions on concentrations of the drug in the bloodstream. Under normal conditions simvastatin is actively metabolized by the CYP3A4 enzyme to produce a bioavailability of around only 5—10%. If present at the same time as the simvastatin, grapefruit juice produces a competitive inhibition of the CYP3A4 enzyme. Less simvastatin is therefore metabolized and there is a consequent increase in the level of simvastatin in the bloodstream. This phenomenon occurs for a limited range of therapeutic drugs, some for the treatment of cardiovascular disease. An unexpectedly high level of simvastatin identified by chemical analysis of a dried blood spot (DBS) sample[13] was shown to result from a case of nonadherence when the volunteer confirmed taking all the drugs at the same time rather than taking the simvastatin at night as prescribed.

Distribution
Once a drug is absorbed it is carried throughout the body and the distribution of a drug from the blood to the tissue depends on the characteristics of the drug, physiologic conditions of the tissue, and the amount of drug bound to plasma proteins. The drug must be carried to the site of action, but will also distribute into muscles and the general biomass, for example, hair, saliva, and nails. Factors affecting drug distribution include protein binding affinity, formation of complexes, and molecular size. There can be serious restrictions to distribution at natural barriers such as the blood—brain barrier.

Metabolism and Pharmacogenetics
The liver is the major site of drug metabolism which is controlled by numerous classes of enzymes identified as Phase 1 and Phase 2 cytochrome P450 enzymes. Due to the large number of enzymes involved, metabolism is a major source of interpatient variability in response to a specific drug. Genetic polymorphisms within these enzymes means a patient may exhibit poor, normal, or ultrarapid drug metabolism.[14] This may affect the benefits from the therapy as, under these circumstances, there will be different drug levels in the blood. Metabolism usually leads to the reduction of the level of active ingredient in the blood but there are instances where it is the metabolite that is more pharmacologically active. Codeine is a good example here as the CYP2D6 metabolism yields the more active morphine to control pain.[15] Ultrarapid metabolizers will produce higher levels of morphine and as a result will experience greater benefit in terms of pain control. Some similar effects are observed for other medicines and patient genetics therefore has an influence on the efficacy of some medicines.

Over-the-counter tests are currently available, from specialist laboratories, universities, and some high street pharmacies in the United States, the European Union, and the United Kingdom, to identify the best medicine for a particular patient. The prescribed medicine for the same illness may therefore be different for individual patients and the choice of drug must be known in order to identify the correct species to monitor if adherence tests are to be carried out.

Excretion

Drug transporters in the biliary tract and kidney facilitate the excretion of unchanged drug and metabolites from the body predominantly in the urine and to a lesser extent in the feces. The kidney is the primary site for excretion and the level of the drug and/or metabolite in the urine will depend on dose, volume of urine produced during the sampling period, and the pH of the urine.[16] The period of collection may extend over several hours and so any data collected from the sample will be time-averaged and therefore cannot produce information for a specific delay time after dosing. There is growing evidence that some low-molecular-weight products of the metabolism of pharmaceutical products can be excreted via the lungs.[17]

Overall Effects of ADME on Drug Levels in Blood

From the previous sections it is evident that there is a dynamic interaction between drug absorption and drug loss by either metabolism and/or excretion. This interaction means that the levels of the drug in the bloodstream will change significantly as a function of time after dosing. This change can be modeled using pharmacokinetics and the expected changes are similar to those shown in Fig. 2.1 for different dose situations. The boundary of the shaded area representing a single dose shows the change in drug concentration in the blood as a function of time after dosing at time zero. Absorption of the drug leads to the fairly rapid rise to a maximum concentration (C_{max}) after a time (t_{max}) at which point the loss of drug by metabolism or excretion becomes dominant and the concentration will decrease up to 24 hours postdosage when more medication is usually taken. For a single dose situation the concentration in the blood will continue to decrease as shown in Fig. 2.1. The trace, post 24 hours, represents the repeat dose situation and shows the trend toward a steady-state minimum concentration after each dosage period. The rate of decrease in the concentration is measured by the half-life ($t_{1/2}$) as indicated on Fig. 2.1. The values of C_{max}, t_{max}, and $t_{1/2}$ have a crucial effect on the level of the drug in the patient's blood.[18]

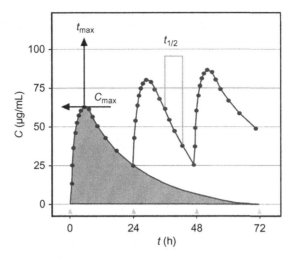

Figure 2.1 General pharmacokinetic plot of drug concentration versus time showing different kinetic parameters.

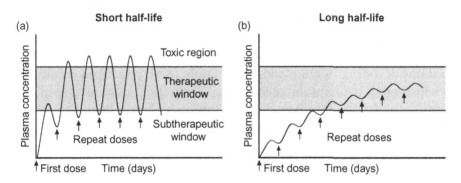

Figure 2.2 Multiple dose pharmacokinetic plot of drug concentration versus time for a short half-life drug (a) and a long half-life drug (b).

Fig. 2.2a reveals that for a drug with a short half-life the plasma drug concentration will oscillate through the "therapeutic window" range whereas for a long half-life (Fig. 2.2b) extra time is taken to reach this level. Flanagan et al.[18] present a mathematical explanation of the levels of drugs in plasma and discuss the potential for assessing adherence to medication by direct monitoring. The changing in plasma levels of the drug or the metabolite, present in a patient's sample, will impose consequent demands on the performance of the analysis used to assess the level of adherence to a particular drug. Tanna and Lawson[19] have discussed the effects of these parameters on the residual levels of selected cardiovascular drugs available for detection in order to demonstrate

Table 2.1 Pharmacokinetic Data for Selected Cardiovascular Drugs[19]				
Drug	Dose Range (mg)	C_{max} (ng/mL)	t_{max} (h)	$t_{1/2}$ (h)
Bisoprolol	2.5/5/10	37–87	1.5–4.0	5–16
Ramipril	2.5/5/10	11–31	2–4	4–6
Amlodipine	5/10	5–7	5–8	35–50
Valsartan	40/80/160	879–3874	2–8	3.5–14
Doxasozin	1/2/4/8/16	17–67	2–7.5	6–21
Simvastatin	10/20/40	5–40	2–3	1.3–2.7
Atenolol	25/50/100	159–377	1.5–6.0	4–11

adherence. Values for these parameters are available in the literature (Table 2.1) and it is not surprising, given the different factors involved with the ADME concept, that there is a range of values for each of the parameters.[18] The data in Table 2.1 shows that both amlodipine and simvastatin have relatively low C_{max} values and that simvastatin reaches this value more quickly and then decays more rapidly due to the shorter half-life. This means that there will be much lower levels of simvastatin in the bloodstream and these must be detected if a false-negative response is to be avoided. These considerations concerning the concentrations of a drug in the bloodstream apply to all prescribed medicines where there is a noninstantaneous, that is, injection, introduction of the drug into the bloodstream. For patients who are adherent to their medication regimens the trough concentrations, indicated on Fig. 2.1 at around 23 hours and again at 47 hours and so on, increase with a move toward a steady-state concentration which will make detection easier.

Physiological factors associated with aging, such as reduced renal and liver function, result in alterations in the pharmacokinetics of drugs, which, in turn increases the risk of adverse drug reactions.[20] In addition, comorbid disease states and the consequent polypharmacy (multiple prescriptions) are common among the elderly, a population in which medication adherence problems are well documented.[21,22] Thus some assessment is vital to guard against adverse drug reactions and to monitor adherence to medication.

Movement of a Drug Through the Body
To directly monitor adherence by identifying a marker of the prescribed drug requires a biosample to be collected from the patient. Current practice is to call the patient into a clinic or family practice where a liquid whole blood sample is collected by a phlebotomist or practice

nurse. This can be an inconvenient and relatively costly process which can be tolerated in more affluent countries but an alternative method is needed particularly where travel and sample transport are more difficult. This approach provides an opportunity to involve the patient experience as the method of sample collection must be acceptable to the patient, particularly if it is to be collected away from a clinic or in a limited "point-of-care" facility. The collection of urine provides the simplest mode of sampling but it provides undefinable time information concerning the distribution of the drug. A liquid blood sample provides fairly precise time-window information but is difficult and expensive to collect and transport and is generally not favored by patients. Alternative, more patient-friendly approaches to the collection of bio-samples can be identified by considering the possible routes of transportation of a drug throughout the body shown in Fig. 2.3.

This diagram is a simplified representation of the pathway taken by a drug or metabolite, within the body, from administration through to excretion. The main blood distribution system is the origin of plasma, the "gold standard" sample while urine has possibly the longest history of use. Collecting capillary blood samples using DBS sampling has become very

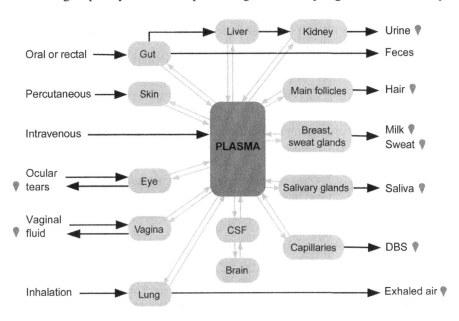

🌡 Indicates a potential sample source

Figure 2.3 Diagram to show routes for drug disposition within the body and potential sample source.

popular following the application of the technique to neonatal screening by Guthrie.[23] This approach is minimally invasive and with modern analytical instrumentation provides the capability to both identify and quantify therapeutic drugs from a single drop of blood. The availability of this detection capability has stimulated the investigation of other less invasive biosamples including saliva, hair, and exhaled breath. The potential of these newer techniques and their limitations are discussed later in this chapter.

BIOLOGICAL SAMPLE MATRICES

In hospitals and clinics, currently the most frequently used human biological sample formats are urine, liquid blood (plasma or serum), and, in special cases, cerebrospinal fluid (CSF). The ease of use and the success of DBS sampling in neonatal screening programs[24] has led to widespread interest in the use of this form of sampling for TDM[25] and ultimately to assess adherence.[13] This success has been derived from two complementary capabilities:

- The ease of sample collection and transport provided by the DBS card
- The ability of analytical equipment to identify and quantify the low levels of target analytes available from the blood spots on the cards

It is only through the increased sophistication and detection capabilities of newer analytical instruments that the micro volume DBS sample (Table 2.2) can provide comparable data to a 1 mL blood sample. This capability has prompted the investigation of alternate less invasive samples including saliva, hair, sweat, tears, vaginal secretions, and exhaled breath. Notional sample volumes and speculative target

Table 2.2 Typical Volumes/Mass of Biological Samples		
Sample	Size	Drug Mass
Urine	100–200 mL	0.1–10 µg
Liquid blood	5–10 mL	10–100 ng
Saliva	0.5–2 mL	0.1–10 ng
Sweat	100–500 µL	0.1–10 ng
CSF	5–8 mL	1–10 ng
DBS	30–100 µL	30–300 fg
Hair	100 mg	30–300 fg
Tears	50–100 µL	30–300 fg

drug amounts contained in these alternative sample formats, based on C_{max} and sample volume, are provided in Table 2.2. Simple non-invasive sampling, carried out at home, backed up by routine analysis of a sample delivered by post could form an acceptable basis for the development of more patient-personalized medical care. The choice of the most appropriate sample would depend on:

- Does the biosample matrix contain the target drug or appropriate metabolite?
- Does the available biosample size contain sufficient target drug or metabolite to be detected?
- Can the biosample provide the therapeutic concentration versus time information?
- Are validated analytical methods available or can they be developed?
- Can the sample be easily collected from the patient?

While these questions have obviously been formatted on the need to appreciate the analytical problems, it is recognized that within any healthcare situation patient collaboration (*willing interaction*) is a crucial factor. Therefore the ease of sample collection must assume a high priority. On this basis, "patient-friendly" sample collection methods should be developed particularly in situations where the nature of the illness dictates the type of sample to be collected.

The rest of this chapter reviews the sampling possibilities likely to facilitate a generally available monitoring service to assess medication adherence based on patient self-sample collection in the home, where possible. There is no discrimination between analyses provided by specialized "pay as you go" laboratories or those provided by the usual healthcare supplier. Inevitably healthcare suppliers will favor approaches in-line with facilities or equipment already in place.

Urine
Patient concerns:
- can be collected at home
- religious, cultural, or ethical issues for some patient groups
- noninvasive
- sometimes difficult to produce/collect sufficient sample especially for the elderly

Healthcare system concerns:
- storage and transport concerns minimal
- must know the appropriate metabolite/s to monitor
- cannot provide therapeutic pharmacokinetic concentration and time information
- sample storage during the collection phase may present problems
- sample authenticity especially in dispute situations, for example, workplace intoxication
- may require expensive and complex analyzers

Urine has been used since ancient times as an assessment of an individual's health but it has become more widespread and publically acceptable, particularly in the United States, following legislation in the 1980s to ensure:

- a drug-free military[26]
- a drug-free federal workplace[27]
- a drug-free department of transportation[28]

The immunoassay test methods developed to deliver the above were based on legal requirements to demonstrate the ability, or otherwise, to perform a designated function. To assess adherence to therapeutic opioid drugs, used in acute pain management, the original test methods have been adapted and modified into a more "patient friendly" approach.[6] When used as a screening test for adherence to therapy the crucial factor in urine drug tests (UDTs) is the "cutoff" point for detection, that is, the point at which adherence is confirmed (see Chapter 4). The majority of literature references citing UDT analyses refer to chronic pain management, but in the United Kingdom urine analysis is also being used as a screening process to assess adherence to cardiovascular drugs in hypertension clinics.[29,30]

Historically, urine has been the favored biosample because relatively large volumes can easily be collected and the quantities collected allow for concentration of samples to meet earlier detection capabilities. This available volume advantage has been virtually eliminated by the increase in sensitivity and specificity of modern instrumental analytical techniques but there is an initial cost premium to be met when purchasing the necessary instrumentation. Cost considerations are important when deciding on the test selected and therefore the necessary pretest work-up of the urine sample. Simple immunoassay or "dip-stick" tests can be readily used with neat or simply diluted urine in point-of-care applications, such as a family clinic, to provide rapid results at minimal costs for this sort of information. The limitations on the data will be those implied by the immunoassay technology which may include reduced sensitivity and specificity when compared with other modern instrumental methods. For immunoassay tests, the detection range will have been preselected depending on the drug type and therefore the "anticipated" urine concentration ranges. The results obtained may be subject to the effects of "cross-sensitivity" and this may be different for different samples.[31]

The steps in a modern instrumental analysis of a urine sample will depend on the nature of the target drug but are likely to include the following:

- initial sample pretreatment
- solid-phase extraction (SPE) or protein precipitation
- quantitative chromatographic analysis
- preparation or purchase of system reference/calibration samples

Modern analytical instruments demand well-supplied laboratory facilities, and the "point-of-care" approach is no longer possible nor is the rapid production of results. Analytical method development can be avoided by simply downloading validated methods for selected therapeutic drugs from analytical suppliers.[32] Based on the example above, the analysis could take anywhere between 1 and 3 hours to be completed but is more likely to be run overnight to allow samples to be batched together to optimize throughput and minimize costs. The "dip-stick" and the full instrumental analysis possibly represent the extremes of adherence testing methodologies, particularly for opioid drugs, and the costs involved may determine which test/s can be used. In the United States, for example, privately funded tests can be purchased from specialist laboratories, whereas in the United Kingdom, only a few selected hospitals receive grants to fund full instrumental analyses for adherence studies. The results from these studies are then used as part of the NHS patient care package.[29,30]

Urine samples are often collected over a period of several hours and therefore do not represent the changing therapeutic drug levels in the patient's blood.

Liquid Blood

Patient concerns:

- needs to attend a clinic to provide the sample
- may be scared of needles
- sample collection is invasive

Healthcare system concerns:

- the "gold standard" method for the quantitative determination of drug levels
- a phlebotomist is required for sample collection
- special storage and transport is needed therefore costly
- liquid blood samples provide sufficient volume for all the required analyses
- cost and complexity of the analyses

Drug measurements are usually made in plasma or serum and care should be taken to determine if there is any preferential distribution of the target drug between these media prior to method selection. Plasma is the supernatant liquid obtained after centrifuging whole blood whereas serum is the liquid remaining after blood has clotted; serum is plasma minus the fibrinogens, the clotting factors.[33] Plasma is currently regarded as the gold standard biosample for drug level determinations. A further complication is the possibility of selectivity within these media. A good example is provided by antiepileptic drugs (AEDs), some of which can bind to serum proteins.[34] While the analytical method measures the total, that is, bound and unbound AEDs in the serum, it is only the unbound that can cross the blood−brain barrier to produce the pharmacological effect. Ideally only the unbound level should be measured, but while the bound/unbound ratio remains fixed the measured total will suffice. Unfortunately there are many instances where this ratio changes at different times within the same patient and also as a result of illness and so methods to separate free from bound components of serum have been developed. These methods include equilibrium dialysis, ultrafiltration, and ultracentrifugation[35−37] but the additional costs involved have led to the development of methods to use saliva for the measurement of free AED.[34]

Dried Blood Spot

Patient concerns:

- minimally invasive
- no need to visit a clinic, can be self-collected at home

- samples can be posted to clinic
- DBS cards can be obtained anonymously prior to sample collection and postal return for analysis. Results are available by post or over the internet

Healthcare system concerns:
- small volume of blood therefore sophisticated instruments required
- high uptake anticipated following success of neonatal screening programs
- initial costs and potential to be automated to speed results and to reduce costs

DBS sample collection has been the basis of neonatal screening for many years and as its name implies the neonatal screening program demanded only a YES/NO result for the disease states targeted. Since the inception of the Guthrie[23] test in 1969 there have been dramatic developments in the performance of several analytical instrument types (see Chapter 3) such that quantification of therapeutic drugs in DBS samples is now accepted as routine.[38] DBS samples are derived from low volume, typically $10-100\,\mu L$ samples of whole blood collected and allowed to dry on an inert matrix. This matrix is usually, but not always, dry cellulose in the form of a card.[39] Blood spot samples from either a finger prick in adults or a heel prick in children are collected on the card either as individual drops or as a known volume collected by pipette prior to spotting onto the card. The first drop of blood is usually discarded and four subsequent drops collected in individually marked regions on the card. The spots are allowed to dry for between 1.5 and 2 hours before being sealed into a small plastic bag with a dessicant. The general approach to a card-based DBS sample is to punch a small disk from the blood spot followed by extraction of the disk with a solvent or buffer solution to provide a solution for further analysis.[40] The size of the punched disk provides the volumetric measure for quantitative analyses and can seriously affect the overall sensitivity of the analysis. More sample, that is, bigger disc, could mean better sensitivity providing the extraction equipment can contain the bigger sample. The actual volume of blood on the punched disk is a function of both the area of the disk and the thickness of the dried blood sample. Blood samples from different people can have different viscosities, related to their hematocrit levels, and will therefore spread differently to form spots of different sizes and thicknesses on the sampling card.[40] Under these circumstances, there will be differences in the volume of blood in a fixed size disc depending on the hematocrit level of the blood. The uncertainty in volume sampled can be eliminated by taking a measured volume of blood and subsequently extracting the entire spot.[41,42] The widespread use of DBS sampling and concern over the "hematocrit problem" has led to the development of several different novel micro-sample collection formats:

- The Mitra[TM] micro-sampling device collects a 10 μL sample directly from the drop of blood on the finger and the volume collected is independent of the patient's hematocrit level (Fig. 2.4a). This device is based on simple volumetric absorptive micro-sampling (VAMS) technology for the collection, transport, storage, and analysis of biological fluids. Four separate sampling devices are supplied in each "clam-shell" pack which can be sealed immediately after sample collection and posted off for analysis as there is no required drying time.[43]
- The HemaSpot[TM] device is designed to aid self-sample collection and needs three drops of blood to saturate the enclosed segmented fan-shaped blood collection media (Fig. 2.4b). After sample collection the device can be immediately sealed, packaged, and despatched for analysis. In the HemaSpot[TM] device the sample for analysis is one of the individual "petals" which is detached from the fan shape prior to extraction.[44]
- The hemaPEN[TM] as its name suggests is like a conventional writing pen but with several barrels (Fig. 2.4c). The blood sample is taken directly from the fingertip avoiding the need to get the drop to the right place on the paper. A second, third, and fourth sample can be separately collected into the integrated storage compartments within the hemaPEN[TM]. This approach ensures precise volume whole blood samples are secured and these can be immediately sent for analysis.[45]
- The Dry Media Spot Slide[46] from TOMTEC is a 6-mm diameter preperforated disk sampling system (Fig. 2.4d) based on either cellulose or a polyester polymer as the collection medium and requiring fixed volume sampling via a 10 μL capillary. Each card has nine spots and all of the sample collected on an individual spot is used when the spot is pushed out for analysis. This approach provides sample quantification and removes concerns related to differences in

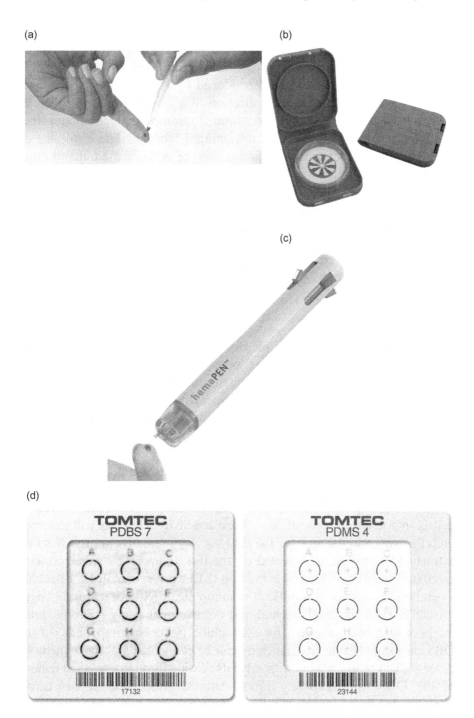

Figure 2.4 (a) The Mitra^{TM} micro-sampling device, (b)the HemaSpot^{TM} micro-sampling device, (c)the hemaPEN^{TM} micro-sampling device, and (d)the TOMTEC dry media spot slide of cellulose and polyester.

hematocrit levels. A range of thicknesses and spot diameters are available for those applications where more sensitivity is required.[46]

- The Noviplex system is included here because the patient sample is a single drop of whole blood produced by either a finger or a heel prick. Noviplex enables the collection of a 2.5-μL aliquot of plasma from a whole blood sample within 3 minutes and eliminates the need for centrifugation and refrigerated storage. The card has a single sample deposition area on top of a laminated membrane stack. The sample is rapidly spread laterally by capillary action in the first membrane layer and then proceeds into the second membrane layer where cells are removed by a combination of adsorption and filtration. Plasma moves down through the membrane matrix to the collection disc at the bottom of the membrane stack. To access the sample the upper layers of the membrane assembly are removed from the card, exposing the plasma filled collection disc. This disc is ready for analysis or transport within 15 minutes.[47]

On receipt of the DBS sample, for quantitative analyses, a decision has to be made concerning the point of inclusion of the internal standard/s (IS) into the process. In most published work the IS has been contained in the extraction solvent.[39] Spraying a solution of the IS onto the card and allowing it to dry prior to the collection of the blood sample may represent a more realistic approach when considering possible analyte interactions with the fibers of the card which may affect extraction efficiency.[48] Simple solvent extraction of the target analyte/s followed by analysis is the least expensive approach. A solid phase extraction cleanup step may be required, prior to analysis, in order to reduce matrix interference problems or increase sensitivity[39] but this will increase both the costs and the time for the analysis. An alternative approach is the extraction of everything collected on the disk followed by either protein precipitation or liquid–liquid extraction (LLE) prior to analysis. LLE has several disadvantages: it is difficult to automate, it requires relatively large solvent levels with environmental and occupational safety concerns, and has poor extraction efficiency for hydrophilic and polar compounds.[39] For DBS card samples the disc punching can be eliminated and the extraction processes can be automated with solvents flowing through a selected region of the DBS using capillary tubes clamped either side of the card. Sophisticated liquid sample handling allows SPE cleanup of the sample to be included in the automated processes.[49]

The alternative DBS sampling formats, cited above, are at a much earlier stage of development and automated sample extraction processes are yet to be developed.

The DBS sample collection can easily be repeated on a daily or more frequent basis to assess the need for dosage changes, for example, during pregnancy[25] in order to optimize healthcare.

Saliva

Patient concerns:
* minimally invasive
* no need to visit a clinic, can be collected at home
* samples can be posted to clinic

Healthcare system concerns:
* small volume of sample therefore sophisticated instruments required
* limited validation of quantitative data
* potential for spurious data from contamination or deliberate dilution

Saliva is a readily available minimally invasive sampling matrix which was used in the 1970s by Danoff and Breimers to study levels of AEDs in patients.[50] The reemergence of this approach is probably due to the increased instrumental detection capabilities now available[31] coupled with the potential, in saliva, to directly measure the therapeutically active free nonprotein-bound drug.[34] While access to saliva samples is relatively easy, the volume available is restricted and may need to be stimulated by a drop of citric acid on the tongue. There are commercial sampling devices available, many of which are nonquantitative and suitable for use in screening applications. Quantitative applications of such devices would require validation of both the methodology and sample stability during transport, prior to data collection.[51−53] Sample collection and transport only requires cleaned sealable tubes and no refrigeration. In the laboratory saliva samples should be centrifuged to produce a clear liquid and any cloudy or discolored samples should be discarded. The fact that saliva samples can be readily contaminated has been demonstrated for both carbamazepine[54] and phenytoin[55] where residues of the tablets held in the mouth for just 5 seconds were still detectable after 2 and 3 hours, respectively. These results might be taken to confirm adherence to these drugs despite such a minimal exposure and noningestion.

One of the undoubted drivers of research into the use of saliva is the ease of sample collection allied to the anticipated cost savings for the health service providers. The cost savings of this form of sample collection were demonstrated for UK offender populations where the costs of the tests were reduced from £500–800 per urine sample to £70–110 for saliva samples used to demonstrate adherence to therapy and abstinence from drugs of abuse.[56] In a Canadian study in 1994 the cost savings of saliva versus blood samples for children were found to be between $1567 and $1822 for every 1000 samples collected.[57]

Saliva samples provide the same opportunities as DBS for repeat sample collection to monitor patient adherence but, whilst the saliva sample is less invasive many more investigations and validated analyses have been reported for the DBS sampling approach.

Hair

Patient concerns:
• minimally invasive
• no need to visit a clinic, can be collected at home
• samples can be posted to the clinic

Healthcare system concerns:
• small volume of sample therefore sophisticated instruments required
• there is no information on the effects of shampoos, dyes, and other treatments on the retention of drugs in hair

Growing hair is continually exposed to the blood system and samples of any drug in circulation will be deposited in the hair shaft. As the hair grows the level of the drug along the shaft will vary in proportion to the level of the drug in the blood. Assuming hair grows at around 1 cm per month analysis of a hair sample can provide information of an average exposure over the long term rather than a day-by-day assessment.

The long digestion period, typically overnight[58,59] needed to release the drugs from the hair structure prior to analysis means that this sample format is not appropriate, as a routine method for a cost-conscious patient healthcare system.

Sweat

Patient concerns:

- elderly patients may not produce sufficient sample
- no need to visit a clinic, can be collected at home
- samples can be posted to clinic

Healthcare system concerns:

- sample collection problems
- small volume of sample therefore sophisticated instruments required
- limited validation of quantitative data

Collection of sufficient sample on an absorbent pad can only be achieved over an extended time period. The time uncertainty in sample collection associated with problems of sample extraction from the pad means that sweat, like hair is not a useful sample format in the current application.[60]

Tears

Patient concerns:

- very uncomfortable for patients to have a collection tube near their eyes
- need to visit a clinic
- not easy to "cry to order"

Healthcare system concerns:

- sample collection problems
- small volume of sample therefore sophisticated instruments required
- limited validation of quantitative data

There appear to be no benefits in the analysis of tear samples and sample collection may well be considered invasive. The concept of chemically inducing patients to cry renders this approach unacceptable as a means of sample generation.[34]

Cerebrospinal Fluid

Patient concerns:

- an invasive procedure
- need to visit a hospital

Healthcare system concerns:

- sample collection costly
- sufficient sample volume but sophisticated instruments required
- sample necessary to provide information on concentrations occurring in the brain

The invasive nature and the costs associated with this procedure would normally prevent any applications in the assessment of adherence to medication except for CNS medication[61] and these samples may be appropriate for Phase 1 clinical trials.[62]

Semen
Patient concerns:
- an invasive procedure
- embarrassing

Healthcare system concerns:
- sufficient sample volume but sophisticated instruments may be required
- usually restricted to HIV/AIDS-related studies

This is not a preferred sample collection method among HIV patients in a comparative study in Kenya, with only 47–48% returning semen samples versus 80% giving saliva samples.[63] This is not a method to adopt for a more general assessment process.

Vaginal Fluid
Patient concerns:
- an invasive procedure
- embarrassing

Healthcare system concerns:
- familiar sample collection process for cervical cancer screening
- sufficient sample volume but sophisticated instruments may be required
- possible adaptation of sampling to HIV/AIDS related studies

A comparative study in Kenya[63] showed that about 52% of the participating female patients provided cervico-vaginal samples during the trials. While being largely disappointing the results from trials of intra-vaginal anti-HIV microbicides, to provide women with a self-controlled means of HIV prevention, have identified unexpected inflammatory effects.[64] This result demonstrates an example of the benefits that can accrue from regular monitoring.

CONCLUSIONS

Some of the suggested biosamples are obviously more patient-friendly than others; urine, saliva, and hair for example would all be easier to collect than a liquid whole blood sample. However there is no value in collecting an "easy" sample if it does not provide the necessary clinical information which is the overriding concern. For information on levels of a drug in blood, a DBS sample, collected at home, might be preferable to a visit to the clinic to provide a blood sample. Alternatively the choice of sample may be specified by the disease. Regardless of its origin the collected biosample will be a complex mixture and some initial sample treatment may be necessary prior to analysis.

The time spent on any initial sample work-up will depend on the level of sophistication of the final analytical assay. The analytical assay may be either:

Direct   where the component of interest is analyzed in the sample as collected, for example, in some immunoassay determinations

Separational   the mixture is separated into its individual components prior to identification and quantification, for example, a gas or liquid chromatograph coupled to a mass spectrometer

Neither of these approaches is necessarily superior and the selection of an assay should be driven by clinical need and not by the technology that might be employed in the assay.

Poor adherence is a worldwide problem, not associated only with the richer nations, and a longer sample work-up time and slower throughput may be a necessary consequence of the purchase of a more affordable analyzer by a less affluent healthcare supplier.

REFERENCES

1. *Pharmaceutical Research and Manufacturers of America.* 2015 biopharmaceutical research industry profile. <www.phrma.org/profiles-reports >; 2015 Accessed 28.10.15.

2. Taylor PJ, Tai C-H, Franklin ME, Pillans PI. The current role of liquid chromatography—tandem mass spectrometry in therapeutic drug monitoring of immunosuppressant and antiretroviral drugs. *Clin Biochem.* 2011;44:14–20.

3. Baughman RP, Meyer KC, Nathanson I, et al. Monitoring of nonsteroidal immunosuppressive drugs in patients with lung disease and lung transplant recipients: American College of Chest Physicians evidence-based clinical practice guidelines. *Chest*. 2012;142(5):e1S–e111S. Available from: http://dx.doi.org/10.1378/chest.12-1044.

4. Warrens AN, Baboolal K, Buist L, et al. Interpreting regulatory authority guidance on immunosuppressive therapy for renal transplantation: a response to the UK's National Institute for Clinical Excellence (NICE). *Clin Nephrol*. 2008;69(2):67–76.

5. Piper C. 10 Popular healthcare provider fraud schemes. <http://www.fraud-magazine.com/article.aspx?id = 4294976280>; 2013 Accessed 24.09.15.

6. Christo PJ, Manchikanti L, Ruan X, et al. Urine drug testing in chronic pain. *Pain Phys*. 2011;14(2):123–143.

7. Kitzmillar JP, Groen DK, Phelps MA, et al. Pharmacogenomic testing: relevance in medical practice. Why drugs work in some patients but not in others. *Cleve Clin J Med*. 2011;78(4):243–257. Available from: http://dx.doi.org/10.3949/ccjm.78a.10145.

8. *Diabetes UK*. TESTING. <https://www.diabetes.org.uk/Guide-to-diabetes/Monitoring/Testing/?gclid = CLWDgsnY6cgCFRJmGwodT_0GcQ>; 2015 Accessed 28.09.15.

9. Staal S, Ungerer M, Floris A, et al. A versatile electrophoresis-based self-test platform. *Electrophoresis*. 2015;36(5):712–721.

10. *BBC News*. Pakistan heart drugs: Lahore death toll reaches 100. <http://www.bbc.co.uk/news/world-asia-16742832>; 2012 Accessed 01.11.15.

11. Reisfield GM, Webb FJ, Bertholf RL, et al. Family physicians' proficiency in urine drug test interpretation. *J Opioid Manag*. 2007;3:333–337.

12. Bair MJ, Krebs EE. Why is urine drug testing not used more often in practice? *Pain Pract*. 2010;10:493–496.

13. Tanna S, Bernier D, Lawson G. LC-HRMS analysis of dried blood spot samples for assessing adherence to cardiovascular medications. *J Bioanal Biomed*. 2015;7:001–005. Available from: http://dx.doi.org/10.4172/1948-593X.1000115.

14. Rogers JF, Nafziger AN, Bertino Jr. JS. Pharmacogenetics affects dosing efficacy and toxicity of cytochrome P450 metabolised drugs. *Am J Med*. 2002;113:746–750.

15. Kirchheiner J, Schmidt H, Tzvetkov M, et al. Pharmacokinetics of codeine and its metabolite morphine in ultra-rapid metabolisers due to CYP2D6 duplication. *Pharmacogenomics*. 2007;J7:257–265.

16. Manno J. Interpretation of urinalysis results. *NIDA Res Monogr*. 1986;73:54–61.

17. Rosenborg S, Meyer MR, Beck O. First report on the pharmacokinetics of tramadol and *O*-desmethyltramadol in exhaled breath compared to plasma and oral fluid after a single oral dose. *Biochem Pharmacol*. 2015;98(3):502–510.

18. Flanagan RJ, Brown NW, Whelpton R. Therapeutic drug monitoring (TDM). *CPD Clin Biochem*. 2008;9(1):3–21.

19. Tanna S, Lawson G. Cardiovascular drug medication adherence assessed by dried blood spot analysis. *J Anal. Bioanal Tech*. 2014;S12:1–3.

20. Sera LC, McPherson ML. Pharmacokinetics and pharmacodynamic changes associated with aging and implications for drug therapy. *Clin Geriatr Med*. 2012;28:273–286.

21. Maher RL, Hanlon J, Hajjar ER. Clinical consequences of polypharmacy in elderly. *Expert Opin Drug Saf*. 2014;13(1):57–65.

22. Olesen C, Harbig P, Blins KM, Barat I, Damsgaard EM. Impact of pharmaceutical care on adherence, hospitalisation and mortality in elderly patients. *Int J Clin Pharm*. 2014;36(1):163–171.

23. Guthrie R. Screening for phenylketonuria. *Triangle*. 1969;9(3):104–109.

24. Chace DH, Spitzer AR, De Jesus VR. Application of dried blood spots in newborn and metabolic screening. In: Li W, Lee MS, eds. *Dried Blood Spots: Applications and Techniques*. New Jersey: Wiley; 2014:53–75.

25. Kulmatycki K, Li W, Xu X, et al. Applications of dried blood spot sampling in clinical pharmacology trials and therapeutic drug monitoring. In: Li W, Lee MS, eds. *Dried Blood Spots: Applications and Techniques*. New Jersey: Wiley; 2014:216–228.

26. Bray RM, Marsden ME, Rachal JV, et al. Drug and alcohol use in the military workplace: findings from the 1988 worldwide survey. *NIDA Res Monogr*. 1990;100:25–43.

27. Reagan R. Executive order 12564. *Fed Register*. September 17, 1986;51(180).

28. Levine MR, Rennie WP. Pre-employment urine drug testing of hospital employees: future questions and review of current literature. *Occup Environ Med*. 2004;61(4):318–324.

29. Lawson AJ, Shipman KE, George S, Dasgupta I. A novel 'dilute-and-shoot' liquid chromatography-tandem mass spectrometry method for the screening of antihypertensive drugs in urine. *J Anal Toxicol*. 2015;40(1):17–27.

30. Tomaszewski M, White C, Patel P, et al. High rates of non-adherence to antihypertensive treatment revealed by high-performance liquid chromatography-tandem mass spectrometry (HP LC-MS/MS) urine analysis. *Heart*. 2014;100:855–861. Available from: http://dx.doi.org/10.1136/heartjnl-2013-305063.

31. Heltsley R, DePriest A, Black DL, et al. Oral fluid drug testing of chronic pain patients. I. Positive prevalence rates of licit and illicit drugs. *J Anal Toxicol*. 2011;35(8):529–540.

32. *Phenomenex*. Applications guide, pain management drugs. <http://www.phenomenex.com/Clinicaldocs>; 2015 Accessed 02.11.15.

33. Tuck MK, Chan DW, Chia D, et al. Standard operating procedures for serum and plasma collection: early detection research network consensus statement standard operating procedure integration working group. *J Proteome Res*. 2009;8(1):113–117. Available from: http://dx.doi.org/10.1021/pr800545q.

34. Patsalos PN, Berry DJ. Therapeutic drug monitoring of antiepileptic drugs by use of saliva. *Ther Drug Monit*. 2013;35(1):4–29. Available from: http://dx.doi.org/10.1097/FTD.0b013e31827c11e7.

35. Kurz H, Trunk H, Weitz B. Evaluation of methods to determine protein-binding of drugs. Equilibrium dialysis, ultrafiltration, ultracentrifugation, gel filtration. *Arzneimittelforschung*. 1977;27(7):1373–1380.

36. Oellerich M, Müller-Vahl H. The EMIT Free Level ultrafiltration technique compared with equilibrium dialysis and ultracentrifugation to determine protein binding of phenytoin. *Clin Pharmacokinet*. 1984;9(suppl 1):61–70.

37. Barré J, Chamouard JM, Houin G, et al. Equilibrium dialysis, ultrafiltration, and ultracentrifugation compared for determining the plasma-protein-binding characteristics of valproic acid. *Clin Chem*. 1985;31(1):60–64.

38. Tanna S, Lawson G. Analytical methods used in conjunction with dried blood spots. *Anal Methods*. 2011;3:1709–1718.

39. Mei J. Dried blood spot sample collection, storage and transportation. In: Li W, Lee MS, eds. *Dried Blood Spots: Applications and Techniques*. New Jersey: Wiley; 2014:21–31.

40. Wong P, James CA. Punching and extraction techniques for dried blood spot analysis. In: Li W, Lee MS, eds. *Dried Blood Spots: Applications and Techniques*. New Jersey: Wiley; 2014:160–167.

41. Li F, Zulkoski J, Fast D, Michael S. Perforated dried blood spots: a novel format for accurate microsampling. *Bioanalysis*. 2011;3(20):2321–2333. Available from: http://dx.doi.org/10.4155/bio.11.219.

42. Youhnovski N, Bergeron A, Furtado M, et al. Pre-cut dried blood spot (PCDBS): an alternative to dried blood spot (DBS) technique to overcome hematocrit impact. *Rapid Commun Mass Spectrom.* 2011;25(19):2951–2958. Available from: http://dx.doi.org/10.1002/rcm.5182.

43. *Neoteryx.* Mitra™ (RUO) microsampling device simplifies collection, transport, and analysis of biological fluids. <http://www.neoteryx.com/>; 2015 Accessed 12.11.15.

44. *Spot on Sciences.* HemaSpot™ blood collection device. The next generation in dried blood spot sample collection. <http://www.spotonsciences.com/hemaspot/>; 2015 Accessed 12.11.15.

45. *Trajan.* hemaPEN™ next generation blood sampling. <http://www.trajanscimed.com/hemapen>; 2014 Accessed 12.11.15.

46. *Tomtec Life Sciences.* <http://www.tomtec.com/assets/dbs_product-brief.pdf>; 2015 Accessed 14.01.16.

47. *Shimadzu Scientific Instruments.* Next generation dried plasma spot technology. <http://www.ssi.shimadzu.com/products/product.cfm?product = noviplex>; 2015 Accessed 12.11.15

48. Zimmer D. Harlan Laboratories. Internal standard application to dried blood spots. <http://www.lc-ms.de/pdf/PDF2011/Zimmer%20ISTD%20Applic.%20%20DBS%20WS%2006.09.11.pdf>; 2011 Accessed 12.11.15.

49. Abu-Rabie P, Spooner N. Dried matrix spot direct analysis: evaluating the robustness of a direct elution technique for use in quantitative bioanalysis. *Bioanalysis.* 2011;3 (24):2769–2781. Available from: http://dx.doi.org/10.4155/bio.11.270.

50. Danof M, Breimer DD. Therapeutic drug monitoring in saliva. *Clin Pharmacokinet.* 1978;3 (1):39–57.

51. Kankirawatana P. Salivary antiepileptic drug levels in Thai children. *J Med Assoc Thai.* 1999;82(1):80–88.

52. Thieme T, Fitchen J, Bartos F, et al. Therapeutic drug monitoring using oral samples collected with the OraSure device. *Ann N Y Acad Sci.* 1993;20(694):337–339.

53. Tennison M, Ali I, Miles MV, et al. Feasibility and acceptance of salivary monitoring of antiepileptic drugs via the US Postal Service. *Ther Drug Monit.* 2004;26(3):295–299.

54. Dickinson RG, Hooper WD, King AR, et al. Fallacious results from measuring salivary carbamazepine concentrations. *Ther Drug Monit.* 1985;7(1):41–45.

55. Ayers GJ, Burnett D. Drug formulation and salivary phenytoin measurements. *Lancet.* 1977;1(8012):656.

56. Bird SM, Pearson G, Strang J. Rationale and cost-efficiency compared for urine or saliva testing and behavioural inquiry among UK offender populations: injectors, arrestees and prisoners. *J Cancer Epidemiol Prev.* 2007;7(1):37–47.

57. Gorodischer R, Burtin P, Hwang P, et al. Saliva versus blood sampling for therapeutic drug monitoring in children: patient and parental preferences and an economic analysis. *Ther Drug Monit.* 1994;16(5):437–443.

58. Tassoni G, Cippitelli M, Ottaviani G, et al. Detection of cannabinoids by ELISA and GC-MS methods in a hair sample previously used to detect other drugs of abuse *J Anal Toxicol.* 2015;:pii: bkv120. Available from: http://dx.doi.org/10.1093/jat/bkv120.

59. De Kesel PM, Lambert WE, Stove CP. An optimized and validated SPE-LC-MS/MS method for the determination of caffeine and paraxanthine in hair. *Talanta.* 2015;144:62–70. Available from: http://dx.doi.org/10.1016/j.talanta.2015.05.084.

60. Kintz P, Tracqui A, Jamey C, et al. Detection of codeine and phenobarbital in sweat collected with a sweat patch. *J Anal Toxicol.* 1996;20:197–201.

61. Cheng Z, Zhang J, Liu H, et al. Central nervous system penetration for small molecule thera-peutic agents does not increase in multiple sclerosis- and Alzheimer's disease-related animal models despite reported blood-brain barrier disruption. *Drug Metab Dispos.* 2010;38 (8):1355–1361. Available from: http://dx.doi.org/10.1124/dmd.110.033324.

62. Cavens L, Ramael S. Cerebrospinal fluid sampling in phase 1 clinical trials: mind over mat-ter. *SGS Life Science Technical Bulletin.* Number 30. <www.sgs.co.uk>; 2009 Accessed 20.10.15.

63. Omosa-Manyonyi G, Park H, Mutua G, et al. Acceptability and feasibility of repeated mucosal specimen collection in clinical trial participants in Kenya. *PLoS One.* 2014;9(10): e110228. Available from: http://dx.doi.org/10.1371/journal.pone.0110228.

64. Smith-McCune K, Chen JC, Greenblatt RM, et al. Unexpected inflammatory effects of intra-vaginal gels (universal placebo gel and nonoxynol-9) on the upper female reproductive tract: a randomized crossover study. *PLoS One.* 2015;10(7):e0129769. Available from: http://dx. doi.org/10.1371/journal.pone.0129769.

CHAPTER 3

Analytical Chemistry Methods for the Assessment of Medication Adherence

INTRODUCTION

Assessment of adherence to prescription drugs implies a direct measurement of the drug or metabolite in a biosample but there are several clinical conditions, for example, diabetes, malaria, HIV, and high blood pressure, where aspects of the disease status itself are measured as an indicator of adherence. While all of these approaches rely on sophisticated instrumental analyses some initial sample work-up is frequently necessary. The analytical methods discussed in this chapter are selected initially to concentrate on the organic and inorganic therapeutic aspects of the medication followed by methods to assess different aspects of the disease status. The final section of this chapter examines the different routes to access the required tests.

SAMPLE PREPARATION

It is important to remember that biological samples are usually a complex mixture with the target analyte often at a low level with respect to the other constituents. While some analyses can be carried out on samples directly ("dilute-and-shoot"), some initial sample cleanup may be used to aid the analytical process by removing interfering contaminants or to prevent excess instrument downtime for cleaning or column changes. It is also necessary to ensure that there is no additional contamination arising from the sample containers. The sensitivity of modern mass spectrometry (MS) systems enables the detection of leachates from some plastics even after very low exposure periods to aqueous solutions. The widespread use of plasticized polyvinyl chloride (PVC) tubes in hospitals provides a ready source of phthalates, for example. Care must also be taken to ensure no loss of the target analyte to the

Analytical Chemistry for Assessing Medication Adherence. DOI: http://dx.doi.org/10.1016/B978-0-12-805463-5.00003-1

sample container. Sample cleanup usually follows one of the two following approaches:

- Removal of unwanted material, for example, by centrifuging or protein precipitation followed by analysis of the supernatant
- Selective extraction of the target analyte/s followed by concentration and analysis

In this area of investigation the samples presented for analysis are liquids and any cleanup procedure must be quantitative where the relationship between the final material for analysis and the original sample is well defined. There are many literature examples of possible cleanup methods applied to samples of different compositions, for example, blood, serum, plasma, urine[1,2], and dried blood spots.[3] The purpose of this part of this chapter is to provide an introduction to each of the cleanup techniques available rather than to detail specific examples. The choice of sample preparation will be dependent on the nature of the primary sample and whether an extract should be analyzed or discarded.

Protein Precipitation

This is a straightforward process in which an organic solvent, usually acetonitrile, methanol, or acetone, is added to samples such as blood, serum, or plasma in order to precipitate the excess protein present. After thorough mixing the sample is centrifuged, to deposit the precipitate on the bottom of the vial, leaving a clear liquid that is removed either for analysis or concentration prior to testing.[4]

Liquid–Liquid Extraction

This is a relatively easy extraction which relies on the different solubility of the analyte between two immiscible liquids. The target analyte passes from the original sample into a more polarity-compatible liquid in which it is more soluble. This process can be used to both extract and concentrate the target analyte/s. Hexane is a good nonpolar solvent which will extract nonpolar drug residues from aqueous media such as urine. Solvent extraction is rarely 100% efficient and repeat extractions (3–5) are required for high recoveries. Relatively large solvent volumes may be used and the process is time-consuming and expensive[2,4]. The additional cost of solvents and subsequent disposal make liquid–liquid extraction processes to be avoided in the provision of service analyses.

Solid-Phase Extraction

This approach to sample preparation uses a solid stationary phase sorbent, held in a cartridge, to clean up and/or concentrate the target analyte, usually by selective retention on the stationary phase. The small mass of stationary phase, a few 100s of milligrams, held in a cartridge means that samples can be flushed through them easily with minimum volumes of solvents. Furthermore, a wide range of sorbent chemistries can be used selectively to retain either the analytes or the unwanted components of the sample.[2] The choice of a particular solid-phase extraction (SPE) cartridge will depend on the chemical nature of the analyte, the sample volume, and whether the analyte is to be retained on the cartridge. For clinical samples the target analyte is usually a nonpolar compound in an aqueous solution and therefore a nonpolar sorbent would be appropriate. To ensure retention of the analyte on the cartridge the typical SPE protocols[5] are as follows:

1. Wetting and conditioning the cartridge: 1−2 mL of methanol is passed through the cartridge to wet all the surfaces, followed by 5 mL of water to condition the surfaces.
2. Sample loading and retention: the sample is passed through the cartridge where the analytes of interest are retained.
3. Wash: any residual unretained sample components are flushed out of the cartridge.
4. Elution: removal of analyte with typically 1−2 mL of solvent. At least a 10-fold increase in concentration is possible at this stage. The sample is collected for subsequent analysis.

In service situations, where many repeat analyses are anticipated, the above protocols can be automated using conventional liquid chromatography (LC) equipment as used for the cleanup of samples from dried blood spots, for example.[6]

INSTRUMENTAL ANALYSES

The last 35 years have seen dramatic developments in new and improved instrumental analytical systems. Modern digital technology has been at the forefront of these new developments with three areas deserving special mention:

- Increase in the speed of signal detection and response of electronic systems

- Development of rapidly accessible bulk data storage capabilities
- Process automation

The increase in the speed of electronic systems was a consequence of the drive for faster computer systems but applied to mass spectrometer systems, for example, this technology increased the measurement capability of time of flight (ToF) systems beyond all expectations of the 1960s–70s to produce the high resolution high mass range instruments of today. Furthermore application of Fourier Transformation theory has benefitted infrared (IR) spectroscopy, nuclear magnetic resonance (NMR) analyses, and latterly MS in the form of the Orbitrap instrument.[7] These instruments are all capable of rapid data generation, which requires bulk storage capabilities and if rapid searching techniques are available, then reference libraries can be constructed and subsequently mined. The National Institute of Science and Technology (NIST)[8] maintains the electron impact (EI) MS database which is in use worldwide to identify unknown compounds in complex mixtures using gas chromatography–mass spectrometry (GC-MS) analyses. NIST is also compiling a LC tandem mass spectrometry (LC-MS/MS) database with information compiled from a range of different instruments and analytical parameters.

The concerns in this chapter are the provision of a service-oriented analytical capability which can take advantage of these modern abilities, particularly automation, to provide:

- rapid analyses with a quick turnaround
- high assay quality—good specificity and sensitivity
- a robust methodology
- affordable costs per analysis, including original equipment purchase
- meaningful data output

Many developments in this area are derived from manufacturers improving current instrumentation to take advantage of a newly emerging market or developing a new research laboratory technique into a service capability. The changes in immunoassay (IA) techniques[9,10] or MS techniques[11] are good examples of the improvement approach, whereas the use of polymerase chain reaction (PCR)[12] analysis is based on the development of a completely new analytical capability.

IMMUNOASSAY (IA) SYSTEMS FOR ORGANIC ACTIVE INGREDIENTS

Unless developed specifically for in-house determinations, IA availability is restricted to kits that are commercially available. Assay performance is therefore well documented, but the kits may be expensive and may not be applicable to specimens other than those identified by the manufacturers.

IA systems depend on the ability of an antibody to selectively combine with the target analyte, the antigen, and for this complex to be measured by the addition of a signaling or label capability. In adherence studies the antigen is either the therapeutic drug or a known metabolite. The antigen–antibody complex binding stage is the selectivity step but there is still the necessity to produce a measurable signal proportional to the level of the complex formed in this binding process. Common assay processes are based either on changing the rate of an enzyme reaction or on measuring the extent of fluorescence polarization.

Enzyme-Based Assays

These assays are usually used with either urine or plasma samples and exploit antigen–antibody binding to influence enzyme activity which can be measured spectrophotometrically. In one approach inactive antibody–enzyme conjugate is added to the test system and the free analyte displaces the enzyme thus increasing the overall enzyme activity in the sample. Measured enzyme activity therefore increases in proportion to the free analyte. An alternative methodology is to use a two-component cloned enzyme donor system. When mixed the two components form the active enzyme. In these assays one component of the enzyme is bonded to the antibody and this component is released as the analyte is added and thus increasing levels of the active enzyme are formed. The measured absorbance signal is proportional to the analyte level.

Fluorescence-Based Polarization Assays

These assays are usually suitable for use with plasma, serum, or urine samples. The labeled analyte will rotate freely in solution and, therefore, when irradiated with polarized light it will emit light which is not polarized. The addition of an antibody produces a construct which cannot rotate as freely and the emitted light retains a greater degree of polarization. An assay will start with the antibody present which produces a high level of polarization. As the free analyte exchanges with the construct the level of polarization decreases and the signal is inversely proportional to the level of the analyte.

To develop detection capabilities at increasingly lower levels manufacturers have used labeling techniques which have included radioactivity, fluorescence, chemiluminesence, and metal chelate complexes.

For automated IA analyses either the antibody or the label may be immobilized to provide the separation necessary for measurement. In automated analyzers micro-well plates are used as sample vials, thus facilitating the washing and reagent pipetting stages. The assay process can be carried out either in the competitive or noncompetitive mode. In the competitive mode there is competition between free analyte in the sample and labeled analyte bound to the antibody which is immobilized. After equilibration and washing the amount of labeled analyte on the surface of the well is measured and is inversely related to the concentration of the target analyte. In the noncompetitive mode two antibodies are required, an immobilized one to bind with the analyte and one, which is labeled, to bind to the antibody—analyte complex. After an incubation period of the sample with the first antibody, the well is washed and a solution of the second antibody is added to react with the immobilized complex. After a second washing the activity of the label bound to the well plate is measured and is proportional to the level of the analyte in the sample.[9,10]

The performance of any IA system is therefore dependent on:

- the selectivity of the antibody and the elimination of interferences[13]
- the capability of the system robotics to track a sample in order to add the necessary levels of reagents and then transfer the sample to the detector at the appropriate time
- the sensitivity of the overall detection system utilized

The antibodies and detection methodologies for a particular antigen (analyte) are often proprietary to a particular manufacturer and may therefore not produce the same detection capabilities on a different instrument. Furthermore an assay "kit" will be tuned to the clinical test requirements and the manufacturers' specifications for a particular assay and will also be subject to specific storage requirements.

Antibodies

Antibodies are proteins generated by animals (cells) in response to the invasion of a foreign molecule, the antigen/analyte, into the body. Because antibodies are developed in response to a particular chemical structure they should respond to it specifically and the analysis of a complex sample (blood or urine) can therefore be achieved without pretreatment.[14] To develop a new IA a new antibody has to be generated and once this has been achieved a complete analysis can be devised. The antibodies can be either monoclonal or polyclonal but for pharmaceutical analyses monoclonal antibodies are advantageous because they have a higher degree of affinity and specificity toward the analyte.[15] Small molecular size and structurally similar therapeutic molecules, for example, the analysis of steroids, pose a significant challenge to the specificity[16] of the IA process. Under these circumstances IA analyses can provide only a screening function indicating the presence of a member of a group of compounds in the test sample and cannot confirm the prescribed compound or the use of an additional medicine of the same compound type.

Instrumentation

IA methods are capable of quantifying a wide variety of compounds from low-molecular-weight drugs to biomarkers of diseases. Automation of the assays is necessary to increase sample throughput. Details of automated IA instruments are tabulated on the CAPTODAY Web site.[17] An assay procedure requires the selection of the correct antibody, the addition of precise volumes of liquids followed by washing with this cycle being repeated prior to measurement. Automated IA analyzers can process multiple samples at the same time, the total number of samples and onboard determinations being dependent on the particular analyzer.[17] The completion time may not be the same for different assays and samples may be introduced into the system at different times and under these circumstances an automated system must have:

- multiple onboard methods with other assays which can be set up as required
- sufficient reagent capacity and accurate dispensing capability for long-term running
- a sample tracking and handling capability to ensure that the correct sample reaches the detector at the correct time for measurement

With instruments of this capability the operator has only to load the samples into individual vials or 96-well plates and then identify the analysis required before loading the samples into the analyzer.

Point-of-Care Testing

Point-of-care testing (POCT) is a simple medical test performed close to the patient where the results are available more quickly than for samples sent to a laboratory. Examples include the use of dipstick monitors, based on IA technology, which are thin plastic strips with several squares of different colors attached along the strip. Each square is used to measure a different component of the sample, usually urine. The entire test strip is dipped into the sample for the stated period and the color changes noted after a fixed development time. Color change reference data is usually portrayed on the side of the container for the test sticks.[18] The chemistry involved in the process leading to the color change is sophisticated and in general is proprietary to the manufacturer. An example of this sophistication is the detection of glucose for diabetes monitoring. Here glucose oxidase liberates hydrogen peroxide from the glucose and this reacts with a potassium iodide chromogen under the action of peroxidase to give a color change from green to brown.[19] The use of dipsticks for screening for drugs or their metabolites is common in many areas, for example, athlete screening, emergency medical care, consulting rooms, and drug detoxification programs. Tests detect the commonly used drugs: cocaine, amphetamines, methamphetamines, marijuana, phencyclidine, barbiturates, benzodiazipines, and opiates.

The main advantages of dipsticks are the convenience, the ease of use, the low cost, and that an assessment of up to 17 drugs can be available within 5−8 minutes depending on the supplier. They can be used in the consulting room or in the home. There are however some limitations to these tests. They are generally not quantitative[18] and the results may vary for different drugs from the same group, for example, the benzodiazipines. There is an insufficient level of specificity toward many of the pain management drugs and IA-based tests in this application are used for screening purposes only.

CHROMATOGRAPHY-BASED METHODS FOR ORGANIC ACTIVE PHARMACEUTICAL INGREDIENTS

Unlike IAs where a degree of specificity is inherent in the use of antibodies or PCR analyses where specificity is derived from the base pair sequences of the analyte itself, chromatographic analysis must develop the necessary specificity from a combination of physical parameters associated with a particular analysis. For all chromatographic analyses the selectivity of a method depends on the separation achieved between the individual components, combined with the capability of the detection system to recognize a particular molecular characteristic. These two signals must be mutually independent of each other; one is time and the other depends on the type of detector being used in the measurements. While this is the ideal situation it must be remembered that adherence problems are now recognized in countries with limited resources and more limited investigations may be the only ones that can be undertaken.

Chromatographic analyses can be divided between the separation of volatile thermally stable compounds using GC and the analysis of nonvolatile materials or thermally sensitive samples by LC. Detectors for chromatographic systems may be either *passive* with no change in detection parameters during an analytical run and will simply indicate an analyte has eluted, or they may be *dynamic* whereby the detection parameters change during the analysis in order to provide specificity toward different compounds as they elute. For a healthcare service provider, where a rapid turnaround is required, the inherently specific analysis represents the optimum because little or no sample pretreatment is required prior to analysis. More extensive initial sample pre-work-up may be tolerated where the final analytical protocol has a lower selectivity toward the target analytes which may be the case where there were financial restrictions on the original equipment purchase. A good example of this mode of investigation of adherence was reported by Bell et al. in a study of the adherence to malaria drugs among patients in Malawi.[20] The chromatographic analysis required multiple solvent extractions with dichloromethane, evaporation to dryness, reconstitution in mobile phase, and analysis by ultraviolet (UV) at 254 nm. Less selective chromatographic methods, with passive detectors, can also be used in situations where the individual dosage level is relatively high, for example, 900 mg allopurinol for gout[21] or 850 mg metformin for diabetes.[22,23]

The principal concern for adherence monitoring is the ability to detect the anticipated levels of the target drugs. This can frequently be achieved with equipment far short of the best.

Chromatography Systems With Passive Detectors

The literature has examples of the use of both gas and liquid chromatographic separations used in the assessment of adherence to medication. The practical applications are discussed in Chapter 4 and the analytical methodology is described below.

GC With Flame Ionization Detection

The basic GC system consists of a gaseous sample in equilibrium between a flowing carrier gas and a heated stationary phase. In instrumental terms there is a heated injector to ensure that gaseous sample molecules are transferred onto the temperature-controlled analytical column where either helium or hydrogen is used as the mobile phase. Technical developments have seen the column change from a one-quarter inch diameter glass column packed with a dry powder to a 0.6-mm diameter capillary system with the stationary phase coated on the inside walls. The other significant change in column technology has been an increase in column length from a maximum of 18 m for the packed column to 100 m or more for the fused silica capillary columns. The column is held in an oven that can be maintained at a fixed temperature or programed typically over the range 30−300°C dependent on the chosen stationary phase. The most widely used detector is the flame ionization detector (FID) which responds to the number of C−H bonds, up to about eight within a compound and thereafter has a fairly uniform response for most organic molecules. Fast detector response and good column technology mean that several hundred individual components can be separated and recorded in a single GC FID run. Identification is derived by matching measured retention times with those for authentic reference analytes[24] and analytical integrity therefore depends on the reproducibility of the retention time which is inherently dependent on the stability of the operating conditions of the GC instrument and a fixed carrier gas flow rate.

For GC analysis the principal factors that affect the separation and detection of the components of a complex mixture include:

Injection volume	Typically between 0.5 and 5 μL
Mobile phase	Helium or hydrogen
Column	Wide range of stationary phases and column lengths
Column temperature	Programmable 30−300°C at preselected rates. Temperature conditions influence both the peak shape and the size of the molecules that will elute from the column
FID	The FID is very sensitive but responding to C−H bonds means it is a good general purpose detector with virtually no selectivity. None of the other available passive GC detectors have been reported in adherence studies to date

Not all molecules are volatile and thermally stable and chemical derivatization to render the molecule volatile is a possible option for many therapeutic drugs[25] so that they can be successfully analyzed by GC techniques. BSTFA (N, O-Bis(trimethylsilyl) trifluoroacetamide) is widely used, in the presence of the catalyst, TMCS (trimethylchlorosilane), to introduce the TMS (trimethylsilyl) group to promote volatility in the target analytes.[25] Derivatization requires that the target drugs be extracted from the biosample using SPE techniques (see earlier) and evaporated to dryness prior to derivatization with BSTFA and 1% TMCS in ethyl acetate. This approach is routinely used in pain management investigations cited in Chapter 4.

Thin Layer and High-Performance Thin Layer Liquid Chromatography
In thin layer chromatography (TLC) the stationary phase is a thin layer of particles a fraction of a millimeter thick, spread on and attached to a rigid base of glass, aluminum, or plastic. The analyte is applied near the bottom of a dry plate and solvents are then used to develop the separation. Plates can be developed in two dimensions using different solvents to improve the separation. Primary advantages of TLC are the low cost combined with the ability to run samples and reference material at the same time.[26] High-performance thin layer liquid chromatography (HPTLC) is a development of the TLC methodology using smaller particles for the stationary phase with a more closely

controlled particle size distribution. For HPTLC, the stationary phase particle size is typically around 5 μm compared with 25 μm for a conventional TLC plate. As a consequence sample sizes have to be smaller for HPTLC, but separations are achieved more rapidly.[26]

Visual detection is acceptable for qualitative analyses but more instrumental methods are required to produce quantitative data. UV, diode array, and fluorescence spectroscopy are the most frequently used methods and direct connection to a mass spectrometer has been reported.[25] HPTLC is one of the most flexible, reliable, and cost-effective separation techniques used in the pharmaceutical industry.[27] Its one drawback may be the need for significant sample work-up prior to analysis to remove potential interfering components from the biosample.

High-Performance Liquid Chromatography With UV Detection
The basic LC system consists of a liquid sample in equilibrium between a flowing liquid mobile phase and a stationary solid phase. The wide applicability of LC methods is derived from the range of different possible mobile phase properties which can be utilized. The choice of mobile phase is not restricted to one or two pure compounds, mixtures can be used and indeed the proportions of the mixture can change during a single analysis. This approach is known as gradient elution, a common approach in biosample analysis. The one requirement is that the individual components of the mobile phase mixture are HPLC grade pure and that there are no UV absorbers present in solution. In instrumental terms an HPLC system consists of a series of reservoirs for the mobile phase constituents connected to a set of pumps which will deliver the liquids according to predetermined values to produce a gradient elution. Because of the high pressure delivered by the pumps samples are injected into a loop injector for subsequent injection into the analytical column. In order to protect the analytical column it is good practice to include a short guard column between the injector and the analytical column. The guard column usually contains the same stationary phase, as a replaceable pellet, as the main column, and is present to prevent unwanted contaminants entering the system. Conventional HPLC columns used for the analysis of organic compounds are reversed phase material packed into 250×4.6-mm stainless steel tubes. Reversed phase material is silica particles with nonpolar groups either $-C_{18}H_{37}$, or $-C_8H_{17}$, or $-CH_3$ chemically bonded to the surface. The octadecyl (C_{18}) group is most frequently used and

abbreviated to ODS or C_{18} on column labels. Under these circumstances the stationary phase is less polar than the mobile phase and nonpolar constituents are retained until the organic level in the mobile phase is increased sufficiently. Improved separation of the components of a complex mixture is achieved by decreasing the size of the silica particles of the stationary phase from, for example, $5-1.3\,\mu m$. Unfortunately this approach increases the resistance to flow through the column and decreased particle size is often accompanied by shorter column lengths; 250 mm down to 50 mm. This change in particle size and the ability to work at higher pressures have been a gradual development and have led to the use of the UPLC and UHPLC acronyms (ultra-pressure or ultra-high-pressure liquid chromatography) to describe systems with particle sizes less than $2.0\,\mu m$. The UV detector used with these systems can operate either at a single fixed wavelength, or at a preselectable wavelength, or at several preset wavelengths in the form of a diode array detector.

For LC analyses the principal factors which can affect the separation and detection of the components of a complex mixture include:

Injection volume	Typically between 1.0 and 100 μL
Mobile phase	Wide range available from a single component to mixtures of components where the individual proportions can vary during the run
Column	Limited range of stationary phases, column lengths restricted to $25-300$ mm range depending on supplier. Particle size ranges $5.0-2.5\,\mu m$ for HPLC and $1.7-1.3\,\mu m$ for UHPLC
Column temperature	Thermostat controlled at fixed temperatures up to $60°C$
UV detector	Very sensitive but has poor discrimination due to wide absorbance bands even for multiwavelength instruments

Chemical derivatization has been used to aid GC analyses by rendering target compounds volatile and therefore detectable and in a similar manner derivatization to add a fluorescent chemical entity to a target molecule could aid detection in HPLC analyses. Under these conditions two different wavelengths must be monitored: excitation at one wavelength and emission at another. The necessity to use two different wavelengths provides a much enhanced degree of selectivity to the analysis. Work using this approach requires considerable sample pre-workup, a direction which has not been followed as the use of more selective dynamic detection systems has been the chosen way forward.

MASS SPECTROMETRY AND CHROMATOGRAPHY

New mass spectrometry technologies have revolutionized the analysis of biosamples. The development of new instrumentation, the tandem mass spectrometer in all its forms, have moved the boundaries of bio-analytical capabilities forward. For any successful combination with a chromatographic system, dynamic detectors must have response speeds fast enough to ensure that multiple measurements can be made within the time span of a single chromatographic event. As each species elutes from the separating system, into the detector, it must be scanned at least $10 \times$ to ensure data clarity.

Mass Spectrometers As Dynamic Detectors For Chromatography Systems

To date, mass spectrometers have proved to be the most successful and widely used dynamic detectors with a range of chromatography options. The different combinations of chromatography and MS reported in the literature relating to adherence studies include:

- GC-MS, GC-tandem MS, or GC-MS/MS
- TLC-MS, LC-MS, LC-MS/MS, and LC−high-resolution mass spectrometry or LC-HRMS

These analytical instrumental combinations are often referred to in the literature simply as hyphenated techniques.

From the technological standpoint, the connection of a high-pressure chromatographic system to an instrument designed to operate under high vacuum presented a range of different challenges. The successful use of MS detection depends on the sample eluting from the column being ionized and transferred into the MS system via a region called the interface. Once ionization is achieved and ions have entered the MS system, data collection and processing are essentially the same whether the detector be a low-resolution linear quadrupole (GC-MS or LC-MS), LC or GC MS/MS system, or a high-resolution instrument (LC-HRMS).

Interfaces for GC-MS Systems

Early GC-MS systems had the problem of the removal of excess mobile phase, either helium or nitrogen carrier gas, from the sample before entry in the vacuum region of the mass spectrometer. Prior to the advent of capillary columns the high gas flows (30 mL/minute) through a packed GC column required the use of an all-glass jet separator to remove the carrier gas via a separately pumped evacuated region prior to the ion source on the MS. The jet separator was maintained above the maximum operating temperature of the column and was extremely fragile. The excess gas flow problem was removed with the advent of the capillary column with gas flows of only 1.0 mL/minute. The jet separator was removed and the flexible end of the column was fed through a heated transfer line, directly into the MS ion source. Sample molecules were ionized by impact with 70 V electrons which produced a wide range of fragment ions with either a positive or a negative charge. These ions were analyzed according to their mass to charge (m/z) ratio. This approach and the use of uprated turbo-molecular pumps produced the benchtop GC-MS systems which are familiar today.

Interfaces for LC-MS Systems

In modern LC-MS systems with a mobile phase flow of between 0.5 and 1.0 mL/minute, ionization is carried out at atmospheric pressure using either an electrospray ionizer (ESI) or an atmospheric pressure chemical ionization (APCI) source. After ionization at atmospheric pressure the ions are transferred through a series of pressure-reducing stages into the mass spectrometer. This is achieved by focusing the ions either through a series of small holes in focusing lenses or through capillary tubes while reducing the pressure with high-capacity pumps. To prevent adiabatic cooling and the formation of clusters as the mobile phase expands under the reduced pressure, the process is carried out in a curtain of heated nitrogen gas.[2]

Electrospray Ionization

This is ionization by charge separation and an electrospray is achieved by forcing liquid through a fine pointed capillary tube, for example, the exit from the LC column, into a high-voltage electric field at atmospheric pressure. The electric field is produced by applying a potential of 3−6 kV, either positive or negative, between the capillary tip and the counter electrode. This field induces charge accumulation in the surface of the liquid at the tip of the capillary which will then form highly charged droplets which move toward the focusing electrode and the entrance to the MS, through the heated curtain gas. As more solvent evaporates from the droplet, under the action of the heated curtain gas, coulombic repulsion forces between like charges overcome the liquid cohesive force and a charged species is emitted. This becomes more frequent as the droplet size decreases. In analyses where ESI is used it is common to modify the mobile phase with up to 0.5% formic acid to provide a source of protons to act as the charge carrier for the ionization process. This addition of a proton to a molecule to produce what is known as a pseudo-molecular ion, the $M + H^+$ species, is a low-energy process in which there is little or no fragmentation of the original molecule. This is not the only ionization process which can happen in an ESI source and species such as $M + Na^+$ can be seen particularly in biosamples. Very large molecules, when subjected to ESI conditions, may become multiply charged to such an extent that the molecular mass of proteins at many tens of thousands can be measured with an MS with an m/z range of only 2000.

While the ESI source is probably the most used in LC-MS applications it has two limitations, namely, the matrix effect and little or no fragment ion production. The matrix effect or ionization suppression or enhancement[2] occurs when other materials from the sample elute into the ESI at the same time as the target analyte and affect the ionization process and hence its efficiency. This is a major problem particularly with clinical samples because of their complexity. This problem can be addressed in three ways:

- Chromatographically improve the separation to prevent co-elution with a consequent increase in sample run time. This is not a favored approach for service providers.
- Selectively remove the interference at the sample preparation stage. Increased sample work-up prior to analysis adds to costs and time per sample analyzed.
- Incorporation of internal standards, usually stable isotope labeled versions of the target analytes, into the analytical process. These will elute at the same time as the target analytes and will therefore experience the same matrix effects and the signal size detected can be used to compensate for the matrix effect observed.

The limited fragmentation as a result of the low-energy (soft) ionization process used means that little structural information can be deduced about the molecular structure. The tandem mass spectrometer is a means of obtaining structural information by forcing the molecular ion to undergo collision-induced fragmentation followed by MS detection of the resultant product ions.

Atmospheric Pressure Chemical Ionization
This is an ionization technique that uses ion–molecule reactions, at atmospheric pressure, to transfer charges originating from the action of a corona discharge on a spray of the mobile phase. The mobile phase and sample from the column are passed through a heated (120°C) capillary tube and are volatilized in the region of a corona discharge electrode where they are ionized. Because this happens at atmospheric pressure the ions are rapidly thermalized by collision and the major ionic products are either the protonated molecular ion $M + H^+$ or adducts from the mobile phase $M + (H_2O)H^+$ or $M + (MeOH)H^+$. APCI systems are usually used with polar or ionic compounds with low (<1500) molecular weights.

Mass Analysis

The mass spectrometers used with both GC-MS and LC-MS systems fall into three broad categories: low-resolution scanning instruments, tandem MS systems, and high-resolution scanning systems. All MS systems measure the mass/charge ratio (m/z) of ionic species and this term is used throughout this work to remind readers that while in general z is assumed to be 1, for many biological species this is not the case particularly where ESI is used. Immunoglobulin gamma is a peptide with a molecular mass of 150 kDa and produces a mass spectrum with m/z values between 4000 and 2000 corresponding to ions with multiple charges. The most intense ions are at m/z 2678.32, 2727.91, and 2630.48 corresponding to 55, 54, and 56 charges, respectively. For a given MS system this approach can be used to calculate the molecular mass of a sample[7] much greater than the quoted m/z range based on $z = 1$. The main characteristics of an MS system are the upper m/z limit, the transmission, and the resolution. The transmission is the ratio of the ion current reaching the detector as a fraction of the ion current leaving the source. The resolution is the ability of the system to separate the signals for two ions with a small mass difference. While the mode of operation of these systems is similar there are some differences which can best be identified by detailing them separately as follows.

Low-Resolution Scanning MS—Linear Quadrupole

The quadrupole mass filter has a mass range of around 3000 with a resolution up to 3000. As its name implies the quadrupole consists of a square array of four parallel circular or ideally hyperbolic rods between 15 and 25 cm in length. Opposite pairs of rods are electrically connected together to either a positive or a negative electrical field with RF and DC components. Ions, from an EI source, entering along the long axis of the device are forced to oscillate away from the center under the influence of the combined fields. Some of these ions will be lost by collision with the rods while others, of heavier mass, will not have moved sufficiently to be lost. The motion of ions in this device is subject to the Mathieu equation,[28] which shows that for certain values of the RF and DC voltages, ions of only one selected mass will have a stable trajectory through the device and reach the detector. A mass spectrum can therefore be generated by scanning the RF and DC voltages over a fixed range while ensuring a fixed ratio between them.[28]

Electrical fields can be rapidly changed and the quadrupole can therefore complete a full mass range scan in a few seconds which is ideally suited to elution time for compounds from capillary GC systems. The downside with this mode of mass analysis is the relatively poor overall transmission of the MS system which results from ions of only one m/z value being stable in the device at any one time. All the other ions are lost as the mass range is scanned.

Commercial GC-MS instruments are relatively small benchtop systems requiring only a single cylinder of carrier gas. For application in adherence assessment, GC-MS analyses often require samples to be derivatized prior to analysis. The main advantage of this approach is the data-rich fingerprint EI mass spectrum which is recorded for each compound eluting from the column. These can be compared with international databases[8] and in combination with a calibrated retention time will provide the necessary specificity for target compound recognition.

Commercial LC-MS systems are similarly sized and have the advantage that, unlike GC-MS systems, there is no necessity for sample derivatization with the ESI source. Because the ionization is a low-energy process the most abundant and possibly the only significant ion formed is usually the $M + H^+$ and so the information-rich fragmentation data is not available. This presents a problem since at m/z 250, for example, there are over 1200 compounds with that molecular mass. In the absence of any corroborating fragmentation data even the correct retention time would not be sufficiently selective. This problem is overcome by tandem MS instruments.

Low-Resolution Tandem Mass Spectrometers—Triple Quadrupole and Ion Traps

The triple quadrupole (qqq) is an example of a tandem mass spectrometer and consists of two quadrupole MS systems in a straight line separated by a quadrupole drift or collision region (Fig. 3.1). There are in fact only two operating mass spectrometers in a qqq system. The tandem mass spectrometer[28] overcomes the problem of poor specificity by forcing the protonated molecular ions, selected by the first MS, to undergo collision-induced fragmentation in a region of relatively high pressure of a nonreactive gas. The collision products are then mass analyzed by the second MS system. This process is shown schematically in Fig. 3.1. Before any analysis can be undertaken the appropriate m/z values for MS1 and MS2 have to be determined. To do this, authentic samples are used to determine the m/z value for the molecular ion (MS1) and these ions are then focused into the collision cell where fragmentation produces the ions characteristic of that particular

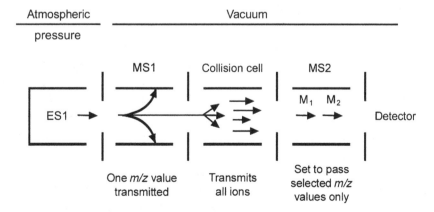

Figure 3.1 Schematic representation of a qqq tandem mass spectrometer (MS/MS) system.

species. The m/z values for these characteristic ions are determined in MS2. The generalized approach for a tandem MS analysis is therefore to select the target molecular ions m/z value in MS1 and set MS2 to monitor the predetermined m/z value/s for the fragment ion/s. One of the m/z values monitored by MS2 is the "quantifier ion" and another is the "qualifier ion" which is used to confirm the correct identity of the species being monitored. This approach is commonly referred to as multiple reaction monitoring (MRM).

The first commercial tandem mass spectrometers were combined with GC systems, and while there was little scope for extensive study of biosamples, the specificity of the technique was demonstrated in the unique identification of a range of tropane alkaloids obtained from plant material with no pre-work-up and infused directly into the MS ion source.[29]

The combination of the ESI source and the tandem MS, in the form of the qqq, has become the widely accepted work-horse instrument in many areas of biosample LC-analysis. This is particularly true in adherence monitoring where the necessity to detect low levels was outlined in Chapter 2. The tandem system is inherently more sensitive than a single quadrupole because more scan time is spent monitoring the target m/z values. In MRM mode MS1 selects relatively few m/z values with respect to the mass range available and therefore the dwell time for an individual m/z value is greater than for the conventional scanning mode and a larger signal is detected. In conventional systems this would be called single ion (m/z) monitoring (SIM).

Ion traps, with ion trajectories based on the Mathieu Equation,[28] can be operated in two modes. Applying the appropriate $RF + DC$ potentials will produce m/z selective ion storage within the volume of the device and if the DC potential is removed ions of all m/z values are stable in the device. This means that ion traps (IT) can also operate as tandem mass spectrometers by alternating between mass selective and nonmass selective modes of operation. Like the linear tandem MS the first action in the trap is to select the target molecular ion to be stable. The DC voltage is removed and the ions are retained in the trap to undergo collision-induced fragmentation as a result of the relatively long residence time. After a suitable period the DC is scanned to identify the m/z values of the product ions. This process can be repeated for one of the product ions, the second set of product ions is identified and so on giving rise to MS^n analyses. When these analyses are carried out in an IT they are referred to as "tandem in time" and in the linear qqq system they are referred to as "tandem in space."[30]

High-Resolution Mass Spectrometers—ToF and Orbitrap
High resolution implies the ability to measure the m/z value accurately to within a few parts per million of the mass and this capability provides an alternate approach to molecular specificity via a calculation or direct measure of the accurate mass, typically to four decimal places, of the target species. A very simple example is the nominal $m/z = 28$ which can be N_2^+ (28.0061) or CO^+ (27.9949) or $C_2H_4^+$ (28.0313) each of which can be separately identified by an HRMS instrument. Not only does this capability provide specificity when identifying therapeutic drugs but it also allows accurate determinations of nonintegral m/z values resulting from multiply-charged large biomolecules, proteins, for example, with more than 50 charges.[7]

Time-of-Flight (ToF) Instruments
In a ToF MS^{26} the m/z value of an ion is determined from the flight time from the source to the detector, a distance d away, under the action of a potential V. The relationship between the flight time and the m/z value is given by:

$$(m/z) = ft^2$$

Where f is a constant derived from the ion velocity and the distance between the source and the detector and t is the flight time. From this equation it can be seen that in principle the ToF has no upper mass limit which is especially useful for soft ionization of large biomolecules. In reality an upper mass limit is arbitrarily fixed by selecting a delay period between pulses of ions from the source to provide a time zero reference point. Furthermore ions are not discarded during the pulsed process and the instrument therefore offers a very high sensitivity. The linear ToF does however have a poor ability to measure m/z accurately. To provide high accuracy m/z determinations some device must be used to correct the small differences in the energy of ions with the same m/z value resulting from small variations in field strength across the ion source. Two approaches are possible:

- Improve ion energy filtering and increase the flight path length but these would lead to reduced signal strength

- Use electrostatic focusing of the ion beam to focus ions of the same *m/z* onto the detector at the same time. This can be achieved using a device called a Reflectron[26] which essentially doubles the path length by focusing the ions back onto a detector which is coplanar with the source.

A Reflectron-based ToF is the current commercially preferred instrument type to minimize the footprint required. Several manufacturers have designed instruments with vertical flight tubes but there are still some restrictions. The accuracy claimed for the determination of *m/z* values using ToF instruments equipped with Reflectron energy focusing is, in general, better than 5 ppm in mass.

Orbitrap

This is a relatively new type of MS instrument[7] which offers high resolution and high mass accuracy based on the orbital trapping of ions between a specially shaped central electrode and an outer barrel-shaped electrode. Ions are transferred into the trap in a series of pulses and ions of all masses are retained within the analytical volume. The ions orbit the central electrode and at the same time oscillate along the central axis. The equations of motion for this device are quite complex[7] and only the axial frequency can be used for mass analysis. The equation of motion in the axial direction is that of a simple harmonic oscillator. Mass analysis can be carried out in two modes:

- Fourier Transform (FT) mode applied to coherent oscillations in the axial direction
- Mass selective instability (MSI) which involves selective ion ejection and detection

The FT mode is used to provide the highest resolution mass spectrum that the Orbitrap can produce, and while this is the principal mode of analysis, the MSI approach will allow MS/MS investigations to be carried out within the Orbitrap.

Data Collection and Manipulation for GC-MS and LC-MS Instruments

Before any data is collected it is axiomatic that the MS instrument performance should be checked and that calibration of *m/z* values has been performed in-line with the manufacturer's instructions using the calibration samples supplied. The supply of reference compounds may also be required to account for variations in instrument calibration over time. Furthermore it is assumed that the instrument is operating under appropriate conditions for the samples to be analyzed and that the anticipated signals will be of sufficient intensity to be measured and will be free from any interferences from other species. It may be routine to run a known sample or sample mix to confirm instrumental performance.

To understand how data is collected, first consider either a simple GC or LC quadrupole system. Once a sample has been injected into the system the quadrupole starts to scan repetitively over the assigned mass range and the number of scans performed becomes a measure of the chromatographic retention time. The sum of the intensities of all the ions detected in a single scan is recorded and is used to represent the level of a compound eluting from the chromatograph. A construct of total ion intensity plotted versus scan number, as shown in Fig. 3.2a, is known as either a total ion chromatogram (TIC) or a reconstructed ion chromatogram (RIC). The mass spectral data, m/z value, and peak intensity are recorded in the MS data system for each scan and the TIC is produced as different compounds elute from the analytical column. This data can be reprocessed at any time. Fig. 3.2a shows four MS scans, one corresponding to the background signal as no compounds elute from the column (low ion intensity) and three showing the different m/z values for separate compounds eluting from the column.

Instead of full scan mode the quadrupole can be used in the SIM mode where the selected m/z values are monitored for extended periods. Under these circumstances the TIC signal output changes, as shown in Fig. 3.2b. The TIC only contains a signal when this particular m/z value is detected. This technique can lead to an increase in sensitivity versus full scan mode but this is dependent on the number of SIM options chosen. The resolution of a single quadrupole is insufficient to provide unambiguous chemical identification of compounds with m/z values greater than ~ 118.[31] But this capability is available from qqq (or tandem instruments) and is derived from MRM events which provide unique fingerprints for the target analyte/s. In Fig. 3.2c the MS data is shown in three groups corresponding to the selection of the primary ion in MS1, a range of ions of different m/z values in the collision cell resulting from collision-induced fragmentation and the quantification and the qualifier

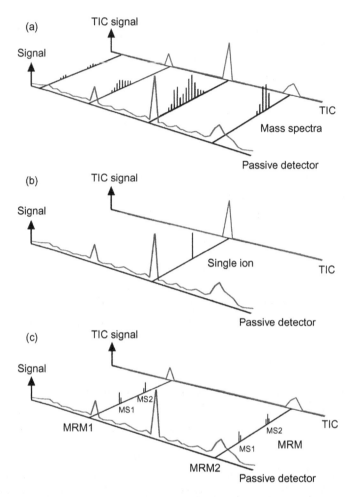

Figure 3.2 (a) Relationship between the chromatographic signal and the TIC, derived from the mass spectral signal, for the analysis of mixture A; (b) effect on the TIC, for mixture A, as a result of selecting SIM for the MS; and (c) effect on the TIC, for mixture A, of selecting two MRM sequences.

product ions monitored by MS2. In this instance the size of the TIC output is determined by the signal from MS2. The shape of the output TIC signal will depend on the dwell time for each MRM event. Under these analytical procedures the data collected relates only to the ions pre-programed into the MRM sequence and all other information is lost as is shown in Fig. 3.2c.

In these applications both the MS analyzers were capable of an *m/z* accuracy of only 50 ppm and were therefore set to monitor nominal or unit mass to charge. High-resolution instruments are able to measure m/z ratios more accurately and it is necessary to understand the different masses that can be determined for the same compound.

Using sulfamethazine $C_{12}H_{14}N_4O_2S$ as an example:[31]

Nominal molecular mass = 278
Average molecular mass = 278.3313—calculated based on all isotopes and natural abundances
Precise molecular mass = 278.0837—calculated for a specific isotopic composition

The ability of both the ToF and the Orbitrap to measure m/z values to four decimal places means that the level of selectivity from the LC component of the analysis can be reduced thus producing a time saving with no consequent loss in the value of the results.[32] Data collected over the total run time of the sample is subsequently reprocessed to give the high-resolution output, a single peak similar to the TIC output shown in Fig. 3.2b. In this instance however all the data from the complete run is archived and can be reprocessed at any time.

Data management is rapidly becoming a major problem. If an individual sample run time is, for example, 5 minutes with a 2-minute turnaround time, then high-resolution mass accurate data can be produced at up to a 1 GB/hour. For big service providers with several MS instruments this can rapidly become several terabytes of data that needs to be stored, handled, and protected.[33,34]

Comparison of MS Systems

The different modes of MS operation result in different performance characteristics, some of which are outlined in Table 3.1. The data is taken from both the literature[35–38] and the manufacturers' application notes and is meant for guidance only as variations in instrumental configuration and selection of different analytes will affect sensitivity information.

A comparison of single quadrupole, ToF, IT, qqq, and QToF systems from the same manufacturer[32] used for the analysis of a fixed set of compounds demonstrated that for repeat analyses of target compounds the optimum instrument was the qqq or tandem MS system. This work also demonstrated that for unknown mixture analysis the QToF was the ultimate instrument with the ability to produce both

Table 3.1 Figures of Merit for a Range of MS Detectors Commonly Used in LC-MS Instruments						
Instrument	Mass Range (m/z)	Resolution/ FWHM	Delta (m)	Scan Rate	Cost ($)	Sensitivity (g)
Linear quadrupole (MS)	40–3000	c.4000	50 ppm	10,000 amu/s	$	$10^{-12}-10^{-15}$
Triple quadrupole (MS/MS)	10–2000	c.2000	50 ppm	Up to 500 MRM/s	$$$	$10^{-11}-10^{-13}$
ToF	20–20,000	>20,000	~3 ppm	40 Hz	$$	$10^{-11}-10^{-12}$
Quadrupole ToF (QToF)	20–20,000	40,000	~3 ppm	20 Hz	$$$$	$10^{-11}-10^{-12}$
Orbitrap	200–4000	15,000–140,000	~1 ppm	18 Hz	$$$$$	10^{-15}

high-resolution full scan fingerprint data and MS/MS fragmentation data to provide structural confirmation of the HRMS data. In a separate publication the potential of the Orbitrap to provide the ability to distinguish between peaks from subintegral m/z values resulting from multiply-charged ions from large bio-molecules was identified.[7]

ANALYTICAL METHODS FOR INORGANIC ACTIVE PHARMACEUTICAL INGREDIENTS

These analyses are concerned with the detection of atoms or ions[39] and not molecular species that were the subject of the previous sections in this chapter. The experiments deal either with atoms that have been released from the biological matrix or with species that are relatively unaffected by the matrix that contains the analyte. In this category the adherence to lithium-based therapies is most frequently assessed.

Flame Photometry

The emission or absorption of light in the UV or visible range is measured at a characteristic wavelength in order to provide a signal proportional to the concentration. Liquid samples are introduced into a sampling stream, fed to the detector, and the output is compared with reference samples. The system can be readily automated, is resilient, and only requires fuel gas.

Ion-Selective Electrodes

Interface potentials, the voltages developed between two phases separated by an interface are the basis of ion-selective electrode measurements. The most frequently encountered ion-selective electrode measures pH which is related specifically to the hydrogen ion concentration. In this case the membrane is a thin, soft glass membrane but for other electrodes the membrane is usually a chemically-modified polymer. The potential across the membrane results from the difference in concentration of a particular ion on either side of the membrane. The potential developed is described by the Nernst equation:

$$E = E^0 + (RT/J) \ln (\text{concentration})$$

To measure lithium ion concentrations in plasma or serum samples a sodium ion-selective electrode can be used connected to a suitable reference electrode. In this format liquid samples can be rapidly tested by immersion of the two electrodes in the sample. The process can be automated and care must be taken to ensure no sample carryover and to ensure calibration checks.[39]

Electrochemical Detection

The Medimate minilab lithium patient self-test at home kit can be used by patients to test a sample of their blood, from a finger prick, to assess adherence to prescribed therapy.[40] The minilab uses moving boundary electrophoresis for sampling from the blood followed by capillary zone electrophoresis for separation of the target analyte and finally conductivity measurements to determine the level present.[40,41] The blood spot is deposited on a disposable cartridge called the lab chip which is then inserted into the multi-reader component. The device detects the presence of the lab chip and begins the analysis which takes 9 minutes after which the concentration is displayed on the screen.

POLYMERASE CHAIN REACTION (PCR) ANALYSIS

DNA is composed of four bases: adenine (A), thymine (T), cytosine (C), and guanine (G); each one attached to a sugar and a phosphate molecule. DNA bases pair up with each other, A with T and C with G to form base pairs. The combination of the base pair with a sugar and a phosphate molecule at each end is called a nucleotide and the nucleotides combine together to give the double helix structure of DNA. The base pairs are essentially holding the two strands of individual bases together to form the intact DNA molecule.

The PCR process[42] takes advantage of the DNA molecule's ability to replicate itself. In the PCR analysis only a short selected region of the original DNA molecule is used. This amplification provides a route to obtain sufficient sample for analysis from very small primary samples and will, if appropriate, minimize inconvenience to patients and reduce transport costs. The first step in the process is to identify the region of the double-stranded DNA to be amplified. This region provides the specificity of the process and must be selected with care. This can usually be achieved by reference to databases of DNA profiles[43] or by in-house determinations. Two short strands of synthetically prepared bases, which are specific to each end of the DNA section to be amplified, are needed to act as primers for the amplification process. The primers are synthesized to ensure the correct half of each base pair is present in the right order to match the region of the target DNA. The number of bases, on the analyte DNA, separating the two primer strands must be consistent with the cycle time of the PCR equipment to be used.

The PCR process is a series of heating and cooling cycles, with each cycle consisting of three separate stages. In stage one a sample of the target DNA is heated to around 95°C to separate the two strands of the DNA by breaking the base pair units. The sample is then allowed to cool slightly to allow the primers to bond with the correct base pair units on the appropriate DNA strand to provide the point for chain extension to begin. Once the primers are in position the temperature is again adjusted to maximize chain extension of the base units attached to the primers. Chain extension, under the action of the enzyme DNA polymerase, can only occur by formation of the appropriate base pair by the incoming base with respect to the base present in the chain at that point. So, for example, if cytosine is present in the sample DNA at this point then only guanine can act as a chain extender. This chain extension process continues on both strands of the target DNA for the time period chosen for this phase of the PCR process. This time period is proportional to the length (number of bases) of the analyte DNA. The end of this time period constitutes the end of the first PCR cycle and at this point the sample is heated to separate the new strands that have formed attached to the primers. The two new strands are duplicate complimentary copies of the strands from the original DNA so the amplification factor at this point is 2. The cycle shown in Fig. 3.3

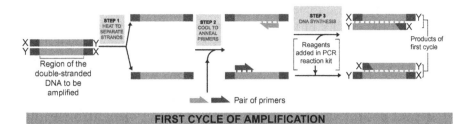

Figure 3.3 Initial stage of the PCR of DNA.

is repeated and new strands are formed on both the original strands and the copies from the previous cycle. The total amplification factor theoretically is 2^n where n is the number of complete cycles.

These steps can be summarized as follows:

1. Prepare samples
2. Heat to 95°C for 30–45 seconds to cleave base pair bonds
3. Cool to 50–60°C for 30 seconds to allow primers to bond with target DNA
4. Heat to 72°C for 30–60 seconds to allow chain extension to develop
5. Repeat steps 2, 3, and 4 between 20 and 40 ×
6. Heat to 72°C for 5 minutes to complete chain extension

This process has two alternative outcomes: if the primers match the target analyte, and become fixed, a large number of duplicate strands of DNA bases will be produced; on the other hand in a no-match situation no extra strands can be formed. After the allotted number of cycles, analysis of the reaction mix is carried out on an electrophoretic gel and the detection of signals at intermediate masses, corresponding to the duplicated strands, confirms the presence of the target analyte in the original sample. This very technically-sophisticated analysis is extremely robust in practice and to set up the PCR process requires only a proprietary sample pack containing the enzymes, DNA bases (A, C, G, or T) and buffers in an incubation tube to which is added the sample and the two primers. The tube is sealed, placed into the chosen PCR instrument,[42] and the process is allowed to run to completion. The analysis of the reaction products is equally robust.

Most of the requirements for good experimental results therefore depend on the quality of the reagents supplied by the instrument manufacturer and for major illnesses such as HIV and malaria there will be well-documented primers available from international suppliers.[42,43] This is important because the specificity of the primers determines the quality of the amplification reaction. The many reports of the use of PCR systems in challenging situations (see Chapter 4) are testament to the ruggedness of the equipment and the approach.

Quantitative PCR

Traditional PCR analysis is nonquantitative providing only a yes/no output in terms of signal amplification. In the PCR process amplification begins slowly, passes through a rapidly increasing period, and then reaches a plateau where all the initial target analyte has been consumed. This plateau is not an indicator of the initial amount of analyte present and for quantitative data another measure is needed. The new approach is a real-time measure of the amplification yield as a function of the number of elapsed cycles. This in turn requires a fluorescent tag attached to the amplified DNA fragments which can be monitored to follow the change from low amplification through the rapid rise and then the plateau. The important point is the number of cycles after which the rapid rise region becomes linear. This point, shown in Fig. 3.4, is known as the threshold value Ct and is proportional to the initial amount of target analyte. The quantitative data required is the

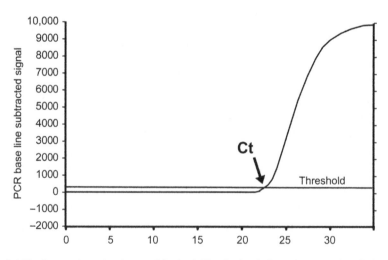

Figure 3.4 This diagram shows the selection of the threshold cycle, Ct, which correlates strongly with the starting level of DNA.

number of the target DNA strands in the sample and this determination will require a series of calibration graphs prepared from mixtures of known DNA concentrations. Calibration traces at different concentrations would be a series of curves and the number of cycles (Ct) after which the rapid increase in numbers becomes linear is determined for each concentration. This data and the appropriate DNA concentrations are used to produce a calibration graph. The test sample is studied in the same way and the number of cycles needed to reach the linear portion of a kinetics plot is determined and this value in conjunction with the calibration graph provides the concentration of the target DNA. There are a number of different ways of processing this information and Pabinger et al.[44] have reviewed 27 open-access software packages for the processing of qPCR data and suggest that more vendors and tools need to use standardized formats to encourage the exchange of data between researchers and users.

TEST FACILITIES

With few exceptions, for example, point-of-care methods or the electronic glucose monitor, the equipment needed to assess adherence to medicines is expensive, complex, and requires specialist attention to maintain system integrity. This implies that test facilities will be maintained in "centers of activity" with the transport of samples or patients/customers into the "center." Sample transport, size, stability, and the need for subsequent destruction and waste removal will all have an impact on the nature of the biosample format selected. Large numbers of urine samples pose significantly different handling problems to a similar number of dried blood spot sample cards. The major items of equipment discussed earlier in this chapter can, in general, only be operated from purpose-built facilities which are usually located in or associated with, one of the following organizations:

1. Hospital laboratories—provide a range of standard clinical tests related to the size of the institution. More specialized tests may be run at a delegated central facility serving a local area. Tests are requested by clinicians and are usually covered by healthcare or special research grant funding. Research tests may be carried out in collaboration with local universities or other specialist research institutions.

2. Public Health laboratories—are usually run by the state or local health departments to monitor the prevalence of selected diseases in the community and also to run rare or unusual tests that clinical laboratories do not offer. Tests are requested by healthcare or government officials.

3. Direct access laboratories—these laboratories allow consumers to purchase tests, from a regulated list, without the need to consult a physician. Such tests are usually bought on the basis of a personal payment and are not usually covered by a health insurance scheme. These laboratories can either be "spin-off" companies from hospital or university research activities or commercial companies set up to provide a limited range of tests as recorded by the American Association for Clinical Chemistry.[45] The laboratories providing these services must also supply consumers with reference range data and some help with understanding their results. Various names have been used for this type of provision including direct access testing, direct to consumer tests, or patient-authorized testing.

All laboratories worldwide providing healthcare, but not research data, are subject to differing forms of government oversight to ensure that samples are handled correctly and that tests are carried out in-line with standard procedures. This requires regular accreditation inspections, education and training programs, proficiency testing, and QC assessments. In the United States the Clinical Laboratory Improvements Amendments legislation of 1988 is used by the Centers for Medicare and Medicaid Services to carry out these checks.[46]

INSTRUMENTATION FOR SERVICE PROVISION

Only two of the approaches previously discussed, namely, IA and MS, can provide the direct determination of drug levels in a bio-fluid and these should therefore be considered as alternative methods for the provision of a monitoring service. Historically MS in clinical laboratories was considered to be time-intensive and costly,[34,36] factors which have inhibited the adoption of the technology particularly when IA techniques were considered good enough. Other barriers to the widespread use of MS systems in hospitals include the cost of the equipment and the training necessary to deal with the inherent complexity of both the LC and MS components of a system. Finally there is the problem of the "front end" of the analytical protocol namely the most appropriate sample to collect and what is the optimum work-up to use? There has been a shift from IA assessments toward MS-based measurements driven by a range of factors:

1. Advances in technology making MS systems able to meet the challenges of the necessary specificity and detection levels.
2. Improved selectivity toward small molecules afforded by LC-MS/MS techniques,[7,38] particularly for immunosuppressant drugs where IA has produced overestimations for both tacrolimus and cyclosporine.[47]
3. Developing new antibodies is a relatively long-term process whereas in tandem MS systems a new species can simply be monitored by identifying three new m/z values (MRM data) for the MS to scan and entering the data into the system.
4. In IA systems, manufacturers may use different measurement technologies for the same target analyte. The commercial reagents are often proprietary and all the necessary assay details may not be available to the end user.[16]
5. There is an increased awareness of interferences in IA determinations[13] and the extent of these effects can vary for different manufacturers' detection kits.

Despite these adverse comments, IA techniques can still offer some significant advantages:

1. Instrumentation is more robust and IA measurements can be carried out in poor laboratory[16] conditions without the need for skilled operators.
2. With automation the assays are high throughput with multiple samples being handled in parallel in the analyzer at any one time. For the MS systems there is only one sample in the LC component of the analyzer and throughput is limited by the retention time required for adequate sample separation.
3. IA systems are intermediately priced with respect to the range and complexity of MS systems available.

For small molecule investigations the advantages, particularly in terms of specificity, appear to lie with the LC-MS/MS systems and while there are limitations on the number of individual samples that can be processed; multiple measurements, using scanned MRM values, can be made on each of these samples. This approach dramatically increases the level of information available from the system and therefore to clinicians. It has been suggested that clinicians accept IA data more readily than clinical biochemists,[47] and despite the limitations, can use the information for diagnostic purposes. An important question is: do clinicians give any thought to how biosample data in a patient's file is collected? This concern is reflected in comments in Chapter 5.

REFERENCES

1. Devanshu S, Rahul M, Annu G, et al. Quantitative bio-analyses by LC-MS/MS. A review. *J Pharm Bio Med Sci.* 2010;7:1–9.

2. Sargent M, ed. *Guide to Achieving Reliable Quantitative LC-MS Measurements.* RSC Analytical Methods Committee; 2013. ISBN 978-0-948926-27-3.

3. Wong P, James CA. Punching and extraction techniques for dried blood spot sample analysis. In: Li W, Lee MS, eds. *Dried Blood Spots: Applications and Techniques.* New Jersey: Wiley; 2014:164–167.

4. Boyd RK, Basic C, Bethem R, eds. *Trace quantitative analysis by mass spectrometry.* Chichester: Wiley; 2008.

5. *Sigma Aldrich.* Solid phase extraction (SPE). <http://www.sigmaaldrich.com/analytical-chromatography/sample-preparation/spe.html>; 2015 Accessed 20.01.15.

6. Ooms JA, Knegt L, Koster EH. Exploration of a new concept for automated dried blood spot analysis using flow-through desorption and online SPE-MS/MS. *Bioanalysis.* 2011;3 (20):2311–2320. Available from: http://dx.doi.org/10.4155/bio.11.214.

7. *CHROMacademy Mass spectrometry.* Fundamental LC-MS, Orbitrap mass analysers. <http://www.chromacademy.com/lms/scol03/Fundamental_LCMS_Interfaces-pdf>; 2015 Accessed 20.01.16.

8. *NIST.* NIST Standard Reference Database 1A. <http://www.nist.gov/srd/nist1a.cfm>; 2015 Accessed 20.01.16.

9. *ImmunoChemistry Technologies.* What is an immunoassay? <www.immunochemistry.com/resource/what-immunoassay/>; 2010 Accessed 20/11/2015.

10. Darwish IA. Immunoassay methods and their applications in pharmaceutical analysis: basic methodology and recent advances. *Int J Biomed Sci.* 2006;2(3):217–235.

11. Shushan R. A review of clinical diagnostic applications of LC tandem mass spectrometry. *Mass Spectrom Rev.* 2010;29(6):930–944.

12. Modern recombinant DNA technology. In: Alberts B, Bray D, Hopkin K, eds. *Essential Cell Biology.* 4th ed. New York, NY: Garland Science; 2013.

13. Tate J, Ward G. Interferences in immunoassay. *Clin Biochem Rev.* 2004;21:105–120.

14. Zhao X, Kottegoda S, Shippy S. Solid-phase immunoassay detection of peptides from complex matrices without a separation. *Analyst.* 2003;128:357–362.

15. Kohler G, Milstein C. Continuous cultures of fused cells secreting antibody of predefined specificity. *Nature.* 1975;256(5517):495–497.

16. Taylor AE, Keevil B, Huhtaniemi IT. Mass spectrometry and immunoassay: how to measure steroid hormones today and tomorrow. *Eur J Endocrinol.* 2015;173(2):D1–12. Available from: http://dx.doi.org/10.1530/EJE-15-0338.

17. *CAP TODAY.* <http://www.captodayonline.com/productguides/>; 2015 Accessed 20.01.16.

18. *Parallax.* Urinalysis test strip color chart. <http://learn.parallax.com/reference/urinalysis-test-strip-color-chart>; 2015 Accessed 5.01.16.

19. Conner JO. Clinitek Status Plus urine dipstick testing with Siemens Multistix Strips Version #2. <http://www.lifebridgehealth.org/uploads/public/documents/lab/nw-clinitek-status-plus-urine-dipstick-testing.pdf>; 2012 Accessed 26.01.16.

20. Bell DJ, Wooton D, Mukaka M, et al. Measurement of adherence, drug concentrations and the effectiveness of artemether-lumifantrine, chlorproguainil-dapsone or sulhadoxine-pyrimethamine in the treatment of uncomplicated malaria in Malawi. *Malar J.* 2009;8:1−9. Available from: http://dx.doi.org/10.1186/1475-2875-8-204.

21. Reinders MK, Nijdam LC, van Roon EN, et al. A simple method for quantification of allo-purinol and oxipurinol in human serum by high-performance liquid chromatography with UV-detection. *J Pharm Biomed Anal.* 2007;45(2):312−317.

22. AbuRuz S, Millership J, McElnay J. Determination of metformin in plasma using a new ion pair solid phase extraction technique and ion pair liquid chromatography. *J Chromatogr B Analyt Technol Biomed Life Sci.* 2003;798(2):203−209.

23. AbuRuz S, Millership J, McElnay J. Dried blood spot liquid chromatography assay for ther-apeutic drug monitoring of metformin. *J Chromatogr B Analyt Technol Biomed Life Sci.* 2006;832(2):202−207.

24. Lewis RJ, Johnson RD, Hattrup RA. Simultaneous analysis of thebaine, 6-MAM and six abused opiates in post-mortem fluids and tissues using Zymark automated solid-phase extrac-tion and gas chromatography-mass spectrometry. *J Chromatogr B Analyt Technol Biomed Life Sci..* 2005;822(1−2):137−145.

25. *Sigma Aldrich.* Derivatization reagents for selective response and detection in complex matrices. <http://www.sigmaaldrich.com/content/dam/sigma-aldrich/migrationresource4/Derivatization%20Rgts%20brochure.pdf>; 2015 Accessed 16.11.15.

26. Rubinson KA, Rubinson JF. *High performance thin layer chromatography. Contemporary Instrumental Analysis.* New Jersey: Prentice Hall Inc; 2000:663−667.

27. Attimarad M, Ahmed KKM, Aldhubaib BE, et al. High-performance thin layer chromatog-raphy: a powerful analytical technique in pharmaceutical drug discovery. *Pharm Methods.* 2011;2(2):71−75.

28. *Waters.* What types of instruments are used? <http://www.waters.com/waters/en_GB/What-Types-of-Instruments-Are-Used%3F/nav.htm?cid = 10090937&locale = en_GB>; 2015 Accessed 21.12.15.

29. Lawson G, Ostah N, Wooley JG. MS/MS studies of tropane alkaloids detection and determi-nation of structure. In: Newton RP, Walton TJ, eds. *Applications of Modern Mass Spectrometry in Plant Science Research.* Oxford: Clarendon Press; 1996:222−228.

30. Johnson JV, Yost RA, Kelly PE, et al. Tandem-in-space and tandem-in-time mass spectrom-etry: triple quadrupoles and quadrupole ion traps. *Anal Chem.* 1990;62(20):2162−2172.

31. *Waters.* Mass accuracy and resolution. <http://www.waters.com/waters/en_GB/Mass-Accuracy-and-Resolution/nav.htm?local>; 2015 Accessed 21.12.15.

32. Tanna S, Bernieh D, Lawson G. LC-HRMS analysis of dried blood spotsamples for asses-sing adherence to cardiovascular medications. *J Bioanal Biomed.* 2015;7(1):001−005. Available from: http://dx.doi.org/10.4172/1948-593X.1000115.

33. *Waters.* Data handling. <http://www.waters.com/waters/en_GB/Data-Handling/nav.htm?locale = en_GB&cid = 10091026>; 2015 Accessed 21.12.15.

34. *Anon.* Education and expense: the barrier to mass spectrometry in clinical laboratories? <http://www.technologynetworks.com/news.aspx?id = 184032>; 2015 Accessed 19.10.15.

35. *Labcompare.* LC-MS instrument. <http://www.labcompare.com/Mass-Spectrometry/155-Liquid-Chromatograph-Mass-Spectrometer-LC-MSInstrument/Compare/?compare=7024678, 6297902,7024679,6297905,6297908&catid = 155>; 2015 Accessed 21.12.15.

36. Zumwalt M, Hughes J, Kilby G, et al. Application note. A comparison of several LC-MS techniques for use in toxicology. <https://www.agilent.com/cs/library/applications/5990-3450EN.pdf>; 2010 Accessed 29.12.15.

37. *CHROMacademy.* Mass spectrometry. Fundamental GC-MS. GC-MS interfaces. <http://chromacademy.com/lms/sco103/Fundamental_GC-MS_Interfaces.pdf>; 2015 Accessed 24.11.15.

38. *Shimadzu.* LC-MS-8060. <http://www.shimadzu.com/an/lcms/8060/index.html>; 2016 Accessed 4.1.16.

39. Rubinson KA, Rubinson JF. *Electrochemical methods. Contemporary Instrumental Analysis.* New Jersey: Prentice Hall Inc; 2000:206−210.

40. *Medimate Minilab.* Lithium in blood. <https://www.medimate.com/en/lithium-blood>; 2014 Accessed 14.01.16

41. Staal S, Ungerer MC, Tellegen M, et al. Therapeutic drug monitoring: lithium patient's self-tests at home. In: *14th Int. Congr. Therapeutic Drug Monitoring & Clin Toxicol. IATDMCT Congress. Rotterdam*; 2015 Poster 329.

42. *Biocompare.* PCR machines. <http://www.biocompare.com/Nucleic-Acid-Electrophoresis/23398-Thermal-Cyclers-Thermocyclers-PCR-Machine/>; 2016 Accessed 04.01.16.

43. *NCBI.* GenBank overview. <http://www.ncbi.nlm.nih.gov/genbank>; 2016 Accessed 4.01.16.

44. Pabinger S, Rodiger S, Kriegner A, et al. A survey of tools for the analysis of quantitative PCR (qPCR) data. *Biomol Detect Quantif.* 2014;1:23−33.

45. *American Association for Clinical Chemistry.* <https://labtestsonline.org/>; 2015 Accessed 10.01.16.

46. *CMS.gov.* <https://www.cms.gov/>; 2016 Accessed 19.01.16.

47. Taylor PJ, Tai C-H, Franklin ME, et al. The current role of liquid chromatography-tandem mass spectrometry in therapeutic drug monitoring of immunosuppressant and antiretroviral drugs. *Clin Biochem.* 2011;44:14−20.

Application of Bioanalytical Methods to Assess Medication Adherence in Clinical Settings

INTRODUCTION

Evidence presented in Chapter 1 suggests that medication adherence problems are observed in all situations where the self-administration of drug therapy is required, regardless of the type of disease, severity of disease, and accessibility of health resources. The self-administration of medications is a developing trend for the treatment of cancer and hepatitis C and is already practiced for a wide range of chronic illnesses including cardiovascular disease (CVD), diabetes, asthma, and depression. Poor adherence to medications is also one of the most important issues to face clinicians treating communicable (infectious) diseases such as human immunodeficiency virus (HIV), acquired immunodeficiency syndrome (AIDS), malaria, and tuberculosis. Perhaps surprisingly non-adherence to immunosuppressants is also prevalent in organ transplant patients. Accurate assessment of adherence is therefore necessary for effective and efficient treatment planning, and for ensuring that changes in health outcomes can be derived from the recommended regimen.[1] The selection of a suitable biosample, such as blood plasma or serum, urine, or saliva, from the patient was discussed in Chapter 2. Methods used to assess medication adherence include direct measures which are objective and confirm the use of medication(s) by assaying them in the biological specimen using the analytical techniques described in Chapter 3. This approach demonstrates the potential of analytical chemistry to address this major healthcare concern. Such direct measures provide quantitative data on the physical presence of medication in a patient's body which can be compared with the anticipated therapeutic window. This objective information can be used by the clinician or healthcare professional to assess the correctness of the dose prescribed, the level of adherence, and the suitability of the medication. While the clinical utility of assaying medications in complex matrices such as biological fluids for assessing

Analytical Chemistry for Assessing Medication Adherence. DOI: http://dx.doi.org/10.1016/B978-0-12-805463-5.00004-3

adherence to prescribed therapy has been demonstrated, there remains the problem of how this might be achieved.

Even for affluent countries where the most advanced instrumentation can be deployed, healthcare costs are such that any adherence assessments will be restricted to major diseases in the first instance. Illnesses such as CVD, cancer, and diabetes would therefore be the expected targets for such an investigation. Malaria and HIV pose major health threats throughout the world but mainly in countries with more challenged health-care facilities and less sophisticated analyses and relying on a more "hands-on approach" would be expected. Sharing of equipment between neighboring countries or shipping samples abroad for analysis could be an option which would define the nature of the sample suitable for this approach.

This chapter looks at disease states with long-term therapies, starting with those indicated by the World Health Organization,[1] in an attempt to determine the extent to which direct assessment of adherence to prescription is being practiced worldwide. The authors live and work in the United Kingdom where the concept of checking on adherence is likened to an act of "Big Brother" (ie, unwanted surveillance) and should not happen. Despite this often-heard statement, the rest of this chapter will demonstrate that such direct assessments not only happen in the United Kingdom but throughout the rest of the world.

CARDIOVASCULAR DISEASE

Cardiovascular disease is one of the biggest killers globally and in the United Kingdom accounted for 28% of all deaths in 2012.[2] Cardiovascular medications comprise the most commonly prescribed medication class in the United States and United Kingdom. The current medical care of patients suffering CVDs uses a combination of cardiovascular therapy drugs from different families. These include antihypertensives (ACE inhibitors, β-blockers, calcium channel blockers, calcium channel openers, angiotensin receptor antagonists, diuretics, vasodilators), hypolipemic drugs (statins), anticoagulants, and antiplatelet drugs. Nonadherence to prescribed medications for the management of hypertension contributes to the burden of adverse outcomes associated with high blood pressure, including myocardial infarction, stroke, and kidney disease. Given the high prevalence and consequences of medication nonadherence, the direct measurement of adherence is a crucial step to ensure that clinicians make an informed clinical decision about treatment and that patients derive the full benefits of the prescribed drug therapy. An objective test of adherence that is simple, quick, and accurate, is therefore needed. Cardiovascular drug analysis in biological fluids constitutes an important challenge for analytical scientists due to the great number of cardiovascular drugs and the frequent use of combined cardiovascular therapy drugs. This is because there is a need for the simultaneous determination of several drugs from different families with different physicochemical properties in biological fluids.[3] Objective studies of the adherence to prescribed cardiovascular pharmacotherapy reported in Table 4.1 have focused

Table 4.1 Bioanalytical Assays to Assess Adherence to Cardiovascular Therapy Drugs				
Analytical Method	Biosample	Analyte(s)	Location	References
HPLC	Urine	Cardiovascular therapy drug—aspirin	United Kingdom	4
HPTLC	Serum	Antihypertensive—propranolol	Chile	5
		Antihypertensive—nifedipine	Chile	6
LC-MS	Urine	Antihypertensives	Germany	7
LC-MS/MS (ion-trap)	Serum	Antihypertensives	United States	8
LC-MS/MS (qqq)	Urine	Antihypertensives	United Kingdom, France,	9–12
	Serum	Antihypertensives	Poland, Czech Rep	13,14
	Serum, plasma	Antihypertensives, statins, diuretics, vasodilators, other cardiovascular therapy drugs	United States	15
	Plasma	Antihypertensives	Germany	16
LC-HRMS (ToF)	Urine	Antihypertensives	Germany	17
	DBS	Antihypertensive—atenolol	United Kingdom	18
		Antihypertensives, statins, diuretics, other cardiovascular therapy drugs	United Kingdom	19,20
LC-HRMS/MS (Orbitrap)	Urine	Cardiovascular therapy drugs	Germany	21
LC-HRMS/MS	Urine, plasma	Antihypertensives	Germany	22
Immunoassay	Serum	Thromboxane B2 (biomarker for aspirin)	United States	23
	Urine	N-acetyl-Ser-Asp-Lys-Pro (biomarker for ramipril)	France	24

on assessing adherence to cardiovascular drug treatment via determination of target drug levels in urine, plasma, serum, and dried blood spot (DBS) samples.

With the exception of Chile the range of countries reporting the investigations belong to the anticipated socioeconomic group for this illness. Where there was a single antihypertensive drug, the analyses have been carried out with high-performance liquid chromatography (HPLC) or high-performance thin layer chromatography (HPTLC).[4–6] HPLC is a powerful separation technique due to its precision, accuracy, and robustness. However, challenges faced with determining combined cardiovascular therapy drugs at very low concentrations has shifted the balance to the development of liquid chromatography–mass spectrometry (LC-MS)-based methods which offer the selectivity and sensitivity required for quantitative analysis of low-level target drugs extracted from complex biological matrices.[7] Triple quadrupole mass spectrometers coupled with liquid chromatography (LC-MS/MS) have been used for a number of applications in this area and remain the primary instrument used in routine quantitative therapeutic drug determinations[8–16] due to the superior specificity and sensitivity they offer. This dominance is now being challenged by high-resolution mass spectrometer (HRMS) in place of the tandem or MS/MS system.[17–22] A study by Patel and coworkers has highlighted that the inclusion of LC-MS/MS urine screening for nonadherence to antihypertensive treatment may be helpful in evaluation of patients with resistant hypertension.[11] Urine and blood samples for LC-MS-based analyses may still require some degree of sample preparation before analysis such as liquid–liquid extraction, protein precipitation, or solid-phase extraction, depending on the characteristics and concentration of the analyte in question. A novel simpler approach with urine samples was a "dilute-and-shoot" method employed to screen a range of antihypertensives.[9] This allowed urine samples to be simply diluted and then directly injected onto the LC-MS/MS system, thus offering a rapid sample preparation time with minimal staff time and reagent costs compared with the standard sample preparation methods.

Urinary immunoassay measurements[23,24] of an endogenous biomarker of ACE inhibition have been used for assessing adherence to an ACE inhibitor (ramipril) in the routine care of diabetic patients.[24] This mode of determination, based on nonradioactive assay, is inexpensive and requires only a random spot urine sample at any moment of the day, with no special requirements for urine sampling. Additionally, it can be performed by nonspecialists. In general, such immunoassays have poorer selectivity than chromatographic-based methods.

Single measurement urine and blood analyses may not fully account for periodicity (both personal and seasonal) of nonadherence to treatment.[25] Here the use of self-collected DBS samples taken at different times could offer a solution as reported by Lawson and Tanna.[18−20] DBS samples, self-collected by the volunteers at different points during drug dosing were analyzed by LC-HRMS in order to confirm the detection of the drugs known to be taken by each volunteer. With this approach personal pharmacokinetic variations may be observed and false negatives resulting from drug−drug interactions may be identified. The advantage of HRMS, in full scan mode, is that all the mass spectral information from the sample is collected. Another consideration is that since data is simultaneously obtained over the whole mass range, it is also possible to directly monitor interfering ions in the sample matrix that could produce ion suppression. Furthermore, HRMS offers excellent mass accuracy and mass resolving power and is now emerging as a feasible alternative to LC-MS/MS for drug quantification studies.

DIABETES

Diabetes mellitus (DM) is a complex heterogeneous endocrine chronic disease and a major clinical and public health concern worldwide where the medication regimen contains many aspects that make adherence problematic. Type 2 diabetes mellitus (T2DM) accounts for the majority of the cases globally. The management of T2DM is complex since many patients are prescribed a very complicated regime of diet, exercise, and oral hypoglycemic agents including several tablets a day. While nonadherence to dietary recommendations has been well described,[26] nonadherence to medication is potentially one of the most serious problems facing diabetes care delivery, particularly in T2DM. Poor control of blood glucose as a consequence of poor diabetes medication adherence can result in poor patient outcomes and the development of complications such as blindness, end-stage renal disease, stroke, coronary artery disease, and peripheral vascular disease thus

Table 4.2 Bioanalytical Assays to Assess Adherence to Oral Hypoglycemic Drugs				
Analytical Method	Biosample	Analyte(s)	Location	References
HPLC	DBS	Metformin	United Kingdom	27
	Plasma	Metformin	United Kingdom	28

adding to the pill burden. Given the seriousness of this chronic illness there are surprisingly few reports of direct measurement assays to assess adherence to oral hypoglycemic drugs (Table 4.2). It is postulated that this may be because diabetic patients are required to routinely monitor their blood glucose levels by a simple fingerprick test. Monitoring adherence is important to improve the control of diabetes in order to prevent unneeded dose adjustment or adding more medication for those who simply have poor adherence.[27]

HPLC assays for the determination of an oral hypoglycemic drug, metformin, in plasma and then in DBS samples are documented and show the relationship between metformin level and metabolic control of diabetes.[27,28] The small volume of the DBS combined with the simplicity of the HPLC analytical technique highlighted the potential of this assay for use in a clinical setting to monitor medication adherence.

CANCER

In the past, most anticancer medications have been given intravenously in a clinic and the question of nonadherence did not occur. Now as more and more oral anticancer medications become available, this situation is rapidly changing and it is envisaged within the next decade, 25−30% of cancer treatments will be delivered in pill form.[29] This change in the treatment paradigm will place greater responsibility on patients to ensure that their medicine is taken correctly. For patients, the convenience of self-administration constitutes one of the major benefits associated with tablet-based therapy. With the rise in availability and increasing use of oral anticancer medications, concerns about the suboptimal use of oral anticancer therapy, its monitoring, and its effects on clinical outcomes are growing for healthcare professionals.[30−32] A major concern that is emerging is the lack of adherence to oral and prolonged cancer therapy. A high degree of adherence, taking medication as recommended, is necessary for the optimal treatment of cancer. Adherence to oral chemotherapy regimens maximizes their effectiveness and minimizes any potential toxicities and is therefore critical for an optimal benefit–risk ratio. However, with some oral anticancer medications, patients may be prone to self-modulating their doses. Without informing their doctor, they may increase their dose due to perceived ineffectiveness or decrease it because of actual or potential side effects. This may then lead the doctor to modify the dose or change to an alternative agent because of apparent nonresponse or unexpected side effects.

For such a major disease the limited amount of data provided in Table 4.3 may be a little surprising but this reflects the new developments away from the clinic and into the home and problems with adherence would be expected to increase in the future. The data presented is material from the open literature and it is impossible to assess the number of tests which may have been carried out by health service providers and private contract laboratories. Nevertheless, there is sufficient information to demonstrate worldwide concern and to provide evidence that adherence to oral medications for cancer varies widely from as little as 14–100% and nonadherence is more pronounced in children and the elderly.[30,33] There is evidence to support a strong recommendation for medication adherence to be assessed routinely and monitored throughout treatment for pediatric cancer patients and this

Table 4.3 Bioanalytical Assays to Assess Adherence to Oral Anticancer Drugs

Analytical Method	Biosample	Analyte(s)	Location	References
PCR	Plasma	BCR-ABL	United Kingdom, Japan, India	34–37
HPLC	DBS	Etoposide	Slovenia	38
LC-MS	Urine	Anastrozole—aromatase inhibitor	United States	39
LC-MS/MS (qqq)	Plasma	Imatinib	United Kingdom, Japan, India	34–37
		Erlotinib	The Netherlands	40
		Anastrozole, letrozole, tamoxifen	Austria	41
		Anastrozole	Austria	42
	DBS	Tamoxifen	Brazil	43
Immunoassay	Serum	Estrone, estradiol	United States	44

would significantly improve outcomes for children and adolescents with cancer.[45] "Overadherence" with oral cancer medications is also reported.[46] This is where too much of a medication is taken by a patient. The issue of nonadherence with oral cancer therapies has been cited by the World Health Organization as the single most important yet modifiable factor that can compromise treatment outcomes.[1]

The treatment of chronic myeloid leukemia has been revolutionized by the development of oral tyrosine kinase inhibitor drugs, such as imatinib, that are very effective in managing the disease when taken correctly. Orally administered hormone-based therapy, including tamoxifen, together with aromatase inhibitors, such as anastrazole, plays an important role in the treatment of breast cancer resulting in substantial reductions in breast cancer recurrence.

As is apparent from Table 4.3, several studies have employed liquid chromatography—tandem mass spectrometry (LC-MS/MS) for the determination of oral anticancer drugs in plasma. For adherence studies on imatinib, coupling these bioanalytical experiments with assessing molecular response levels by PCR techniques was used[34–37] and work of this type confirms the complementary value of the different approaches and can validate the use of the PCR approach in other applications. Nonadherence to the prescribed anticancer drug therapy may be the main reason for poor molecular responses, and Oberguggenberger and coworkers compared the differences between the reported adherence rates for anazatrazole using four different approaches.[42] Three of these approaches were indirect methods and these were compared with a direct method of measuring serum drug concentrations. They concluded that the reported adherence rates depended on the assessment method used, and furthermore, the determination of drug serum concentrations at regular intervals may provide a more meaningful measure of patients' adherence. Tamoxifen and its metabolites have been determined using LC-MS/MS in DBS samples,[43] and this form of sampling and analysis has enormous potential to monitor medication adherence as postulated by Tanna and Lawson.[47]

HIV/AIDS

HIV is a major global public health issue and in 2014[48] there were approximately 37 million people living with HIV globally with

Table 4.4 Bioanalytical Assays to Assess Adherence to Antiretroviral Drugs				
Analytical Method	Biosample	Analyte(s)	Location	References
PCR/ qPCR	Plasma	HIV RNA	United States	49
		HIV RNA	Cambodia → Paris	50
	Plasma, DBS	HIV RNA	Spain, India	51,52
		HIV RNA	Tanzania → Spain	53
	Fingerprick blood collection device	HIV RNA	South Africa	54
TLC	Saliva	Antiretroviral therapy drug—nevirapine	Tanzania	55
	Hair	Antiretroviral therapy drug—nevirapine	United States	56
HPLC	Plasma	Protease inhibitors	United States, The Netherlands, Spain	57–59
		Nonnucleoside reverse transcriptase inhibitors and protease inhibitors	The Netherlands	60
	Saliva, plasma	Antiretroviral therapy drug—indinavir	The Netherlands	61
	DBS	Antiretroviral therapy drug—atazanavir	United States	62
LC-MS	Plasma	Antiretroviral therapy drugs—efavirenz, nevirapine, lamivudine	Malaysia	63
LC-MS/ MS (qqq)	Plasma	Antiretroviral therapy drugs	United States, China, United Kingdom	64–67
	DBS, plasma	Antiretroviral therapy drugs	Germany, The Netherlands	68,69
	Hair, plasma	Antiretroviral therapy drugs	Kenya, Uganda → United States	70
	Hair	Antiretroviral therapy drugs	United Kingdom, Uganda → United States, Kenya → United States	71–74
	Saliva, plasma	Antiretroviral therapy drugs—zidovudine, nevirapine, lamivudine	Rwanda → Luxembourg	75

almost 26 million sufferers in Sub-Saharan Africa alone. The extent of the situation is represented by the data in Table 4.4 which demonstrates:

- The potential of virtually all the analytical techniques discussed in Chapter 3

- The need to use different biological samples dependent on where the sample was collected and where it was to be analyzed (see Chapter 2)
- The collaborative nature of the investigations with samples being transferred either to different countries in the European Union or to the United States for analysis as indicated by →.

The HIV targets the immune system and weakens people's defense against infections and some types of cancer. As the virus destroys and impairs the function of immune cells, infected individuals gradually become immunodeficient, consequences for which are increased susceptibility to a wide range of infections and diseases that people with healthy immune systems can fight off. The most advanced stage of HIV infection is AIDS, which can take from 2 to 15 years to develop depending on the individual. There is no cure for HIV infection; however, effective treatment with antiretroviral therapy (ART) drugs can control the virus so that people with HIV can live longer, healthier lives. People on ART take a combination of HIV medicines (called an HIV regimen) every day.[76] Meticulous adherence to ART drugs is vital since it prevents the virus from multiplying and destroying the immune system and is therefore essential to achieve an optimal clinical and virological response. Some studies have suggested adherence rates of at least 90–95% are required for optimal viral suppression.[77] Poor adherence to an HIV regimen allows HIV to destroy the immune system and also increases the risk of drug resistance and HIV treatment failure. Adherence to an HIV regimen can be difficult for several reasons which have been highlighted in Chapter 1. Adherence to ART pharmacotherapy can be assessed by monitoring the viral load, using PCR, once medication has been initiated. Conventional assessment of viral load in HIV patients is based on a measure of the number of HIV RNA units/mL of plasma or serum. A decrease in the viral load as medication progresses should be observed.[78] Any subsequent increase in the viral load will be indicative of either nonadherence to medication or drug interaction or malabsorption.[79] Alternatively, it can be done by direct measurement of the target ART drug(s) in the biological sample. The PCR assessment is not only less costly than the LC-MS/MS technique but also has a more robust "kit-based" methodology, and furthermore it can also be used to identify changes in the nature of the virus itself.

Antiretroviral drug concentrations are routinely measured in plasma at a single time point obtained by venepuncture. This form of sampling has also been used to ascertain viral load measurements. However, this sampling strategy may pose several disadvantages[80] for assessing adherence to ART especially in a resource-limited setting:

- Difficult to carry out venous sampling in some populations, for example, pediatric patients
- Sampling carried out by phlebotomist or trained personnel at hospital or clinic
- Single drug concentration measurements may not reflect drug exposure during a longer period, because it is only relates adherence to the last intake

For these reasons, other, more patient-friendly, sampling methods have been investigated for assessing adherence to ART including DBS, saliva, and hair and in some cases replicated with conventional plasma measurements.

The analysis of DBS samples is being used to monitor HIV viral loads using PCR techniques. Conventional assessment of viral load in HIV patients is based on a measure of the number of HIV RNA units/ mL of plasma and replacing the liquid with DBS offers many potential cost and efficiency savings.[47] The principal areas of concern were the detection capabilities of the absolute viral load, which depended on the amplification factor of the system and the initial quantification of the blood sample collected. Correlations have been demonstrated between DBS and plasma samples for the assessment of the HIV viral load in a resource-limited setting in rural Tanzania by Johannessen et al.[53] and highlight the advantages of using DBS for assessing adherence in this setting. This investigation constituted a "real life" test of all the advantages highlighted for the DBS approach highlighted in Chapter 2. The priority of this work was to examine if this form of blood sampling was robust enough to use under basic field conditions. The overall conclusion was that DBS sampling systems performed well in monitoring the viral loads for patients who had previously received ART treatment. Adherence to ART pharmacotherapy has also been conducted via the direct measurement of the target drugs in DBS samples by HPLC and LC-MS/MS and investigations confirmed the close relationship between the DBS and plasma concentrations of the target ART drugs.

ART drug concentrations in saliva and hair samples, determined using TLC, HPLC, and LC-MS/MS, have been used to provide noninvasive tools to assess adherence to ART therapy.[55–75] The findings showed that saliva and hair concentrations are correlated with plasma concentrations, and that some assessment of adherence by measuring the drug salivary or hair level is possible. Furthermore, the benefits of using TLC to assess adherence to nevirapine in Tanzania have been highlighted,[55] where more expensive analytical methods such as HPLC and LC-MS-based methods are not available.

Patient hair samples can be obtained noninvasively and ART drug levels in hair can provide retrospective information regarding drug exposure and thus can be used to assess adherence to ART pharmacotherapy. Furthermore, hair as a biological matrix has considerable stability for long periods of time.[81] Drug incorporation into hair is a complex process that is not yet fully understood. The major pathway for drug delivery into hair is by the bloodstream into the hair follicle. Drugs from external sources or drugs excreted in sebum or sweat may also be incorporated or externally contaminate the hair. Hair drug levels depend not only on blood concentrations but also on the physicochemical properties of the drug. Lipophilic and basic drugs such as the ART drugs accumulate well in hair.[80]

Saliva is also a patient-friendly alternative matrix to assess adherence to ART drugs. Saliva can be obtained by nonstimulated methods (direct collection into a tube) or by means of stimulated methods (with a cotton swab containing citric acid). The transport of ART drugs into saliva is mainly by passive diffusion since only drugs that are not extensively protein bound penetrate well into saliva. Drug concentrations in saliva may therefore be useful for predicting the levels of unbound plasma ART drugs. Most ART drugs are highly protein bound, resulting in poor salivary penetration.[80]

Proper sample pretreatment is essential for drug analysis in DBS, hair, and saliva matrices and a specific assay should be developed and validated for application in a clinical real-world setting.

HEPATITIS C

The hepatitis C virus (HCV) is the most common blood-borne chronic infection affecting 130–150 million people globally. The most frequent routes of infection are through needle sharing by individuals with an intravenous drug addiction; inadequate sterilization of medical equipment; and the transfusion of unscreened (contaminated) blood and blood products.[82] Untreated chronic HCV infection is the leading cause of chronic liver disease, including cirrhosis and cancer and the most common cause of liver failure and transplantation. The treatment and management of chronic HCV has evolved in recent times from pegylated interferon and ribavirin, which required weekly injections to the new oral self-administered direct antiviral agent (DAA) drugs which entered the market recently in 2013.[83] Adherence is a crucial factor for the successful treatment of HCV infection; however, given the recent introduction of the new DAA drugs, it is not surprising that very few studies have examined adherence to HCV therapy using direct methods of assessments in real-world settings. DBS testing for blood-borne viruses was introduced in England, Scotland, and Wales[84] a few years ago as an alternative to venepuncture following the reports of Hickman et al.[85] and Crane et al.[86] on the positive value of DBS testing in prison environments and for injection drug users. This form of blood sampling is continuing to contribute to the uptake of HCV testing with numbers tested by DBS in England increased by 23% between 2013 and 2014 and 18% testing positive in 2014 by this sampling method.[84] Tait et al.[87] also confirmed the effectiveness of the DBS sampling approach in reaching people who injected drugs in order to identify blood-borne viruses and subsequently assess the adherence to follow-up medication. In this study, 100% adherence was reported when the DBS samples were collected at a needle exchange center and 83.3% adherence from DBS samples collected at a drug treatment center. The DBS sampling platform clearly has the potential to assess medication adherence in a nonclinical setting, such as a prison environment, and thus could be a vital tool in controlling the global HCV epidemic.

PAIN MANAGEMENT

Pharmacotherapy is essential to the clinical management of most chronic nonmalignant pain conditions; however, therapeutic use, overuse, abuse, and diversion of controlled substance prescription (scheduled) drugs such as opioids and benzodiazepines remain critical challenges for clinicians.[88] It is therefore crucial to practice adherence monitoring of pain management drugs. Urine drug testing

Table 4.5 Bioanalytical Assays to Assess Adherence to Pain Management Drugs				
Analytical Method	**Biosample**	**Analyte(s)**	**Location**	**References**
TLC	Urine	Buprenorphine—naloxone	India	89
TLC, GC-MS	Urine	Benzodiazepines	Germany	90
GC-MS	Urine	Opioids	United States	91–93
LC-MS	Plasma	Opioid—buprenorphine	France	94
LC-HRMS (ToF)	Urine	Opioids, benzodiazepines	United States	95
LC-MS/MS (qqq)	Urine	Opioids	Germany, United States	96–101
		Benzodiazepines	United States	102–104
		Opioids, benzodiazepines	United States	105,106
	Urine, saliva	Opioids	United States	107
		Opioids, benzodiazepines	United States	108
LC-MS/MS, GC-MS	Urine	Opioids	United States	109
LC-MS/MS, LC-MS, GC-MS	Urine	Opioids	United States	110
Immunoassay	Urine	Opioids, benzodiazepines	United States	95,100,102,105,107,108

is a widely used method for adherence monitoring but Table 4.5 shows that there are new developments in the use of both saliva and plasma-based methods.

The types of assays that are being used by clinicians to assess adherence to pain management therapy can be categorized as:

- *Screening (qualitative) assay*
  - May identify drugs and or drug metabolites with variable specificity, often only by drug class
- *Confirmatory (quantitative) assay*
  - Identify and quantify the drug and/or drug metabolites with high specificity

The screening assays are immunoassays that use antibodies directed against a drug or metabolite. These immunoassays may be in a point-of-care test format so that the assay can be carried out in the clinic or they can be laboratory-based commercial immunoassays.

Gas chromatography—mass spectrometry (GC-MS) analysis was the accepted approach to the determination of pain management drug levels but this technique required sample prederivatization and is now being replaced by LC-MS/MS methods because samples can be analyzed directly without resort to chemical derivatization.

This change in approach to assessing adherence to prescription painkillers is confirmed by the data in Table 4.5 which includes urine screening by immunoassays followed by targeted confirmations with LC-MS/MS, liquid chromatography high-resolution mass spectrometry (LC-HRMS), gas chromatography, or GC-MS. Urine is the preferred biosample for these assays because of the long detection window available as a result of the pharmacokinetics for the drug metabolites. Immunoassay-based urine drug tests can be qualitative or quantitative, however they suffer from poor specificity as reflected by the incidence of false-positive and false-negative results. A major limitation of immunoassays is poor sensitivity or a relatively high minimum detection level, versus LC-MS/MS methods for example. The work carried out by Mikel et al. demonstrates how a patient may test "negative" on a urine immunoassay which has a higher cutoff or limit of detection in comparison to an LC-MS/MS assay.[106] In this study the urine opiate immunoassay had a cutoff of 300 ng/mL while the LC-MS/MS assay had a cutoff or limit of detection of 50 ng/mL. For the hydromorphone results in this study, approximately 69% of patients who were prescribed and taking hydromorphone tested "positive" or had detectable drug concentrations from the LC-MS/MS assay, but tested "negative" on the immunoassay due to the higher cutoff and poor cross-reactivity with the antibody used in the immunoassay. Saliva tests to monitor adherence to prescribed painkiller drugs produced results comparable to urine tests with some minor differences in detection rates for different drug classes. This may be attributed to factors affecting the accumulation and diffusion of drugs in saliva such as protein binding, rate of oral fluid production, and the chemical properties of the drug. Generally, basic drugs appear in higher concentration in the saliva than those of an acidic nature due to "ion-trapping" caused by the lower pH of saliva compared to blood. The opioid pain management drugs display high saliva:plasma (S:P) ratios, indicating they accumulate well into saliva and can easily be detected. Drugs which are strongly protein bound (eg, benzodiazepines) generally do not appear in high concentration in saliva.[111]

Within the United States, there are many different private resources providing drug testing services[112,113] for pain control drugs and the existence of these many facilities is an indicator of the number of tests being carried out. Alongside demonstrating adherence, these tests can be used to identify unauthorized drug use[109] or the nonuse and subsequent onward selling.[114]

Serum, plasma, and DBS assays for pain management drugs may offer advantages over urine because the drug level in these biological matrices reflects the patient's clinical state at the time of sample collection. The levels could be used for titrating dose, determining steady-state drug concentrations or determining the drug/metabolite levels in cases of suspected toxicity.[115]

SCHIZOPHRENIA

Schizophrenia can be a severe mental illness as it impairs important aspects of human emotion and cognition that are vital for living in a complex social setting. Antipsychotic medications are the primary treatment option for the management of schizophrenia as well as for relapse prevention in long-term maintenance treatment. There are several factors that make nonadherence to antipsychotic medication a major problem in the management of schizophrenia. These include a lack of illness awareness, social isolation, stigma, comorbid substance misuse, depression, cognitive impairment, and the increasing fragmentation of health services in many countries.[116] The adherence rates to prescribed drug therapy in schizophrenia are divergent and range from 11% to 80%, with an overall rate estimated to be about 50%. The actual rate of adherence may be even lower as the estimates do not account for patients who refuse treatment or drop out of follow-up studies.[117] Nonadherence with antipsychotic medications increases the risk of relapse. Nonadherent patients are also at higher risk of dangerous behavior, suicide, arrest, violence, drug and alcohol consumption, poor mental performance, and low satisfaction with life.[118] Thus, there is a pressing need to objectively assess adherence to antipsychotic medications. In Table 4.6 a number of objective studies which have the potential to offer proof that the patient has taken their antipsychotic medication, that is, detection of the drug in a biological sample, are summarized.

Various analytical methods have been used for the determination of antipsychotic drugs in biological samples in adherence studies. The

Table 4.6 Bioanalytical Assays to Assess Adherence to Antipsychotic Drugs				
Analytical Method	Biosample	Analyte(s)	Location	References
HPLC	Plasma	Serotonin re-uptake inhibitor—fluvoxamine and antipsychotic—risperidone	United States	119
	Plasma, serum	Antipsychotics	Italy	120
		Serotonin re-uptake inhibitors and antipsychotics	Germany	121
HPTLC	Serum	Antipsychotic—clozapine	Chile	122
HPLC, LC-MS, LC-MS/MS	Serum	Antipsychotics	Czech Republic	123
LC-MS/MS (qqq)	Urine	Antipsychotics—risperidone, quetiapine, olanzapine, haloperidol	United States	124

earlier methods were HPLC-based assays for plasma/serum samples, while the most recent assay used LC-MS/MS to quantitatively determine antipsychotics in human urine samples.

After a given oral dose of an antipsychotic drug, the resulting blood concentrations of these drugs are highly variable among individuals. This is because antipsychotics are extensively metabolized in the liver by the cytochrome P450 enzymes. Therefore, the concentration of these drugs in blood is very low. The clinical monitoring of patients can significantly improve the knowledge of pharmacological interactions among different antipsychotic drugs, as well as improve the adherence of these drugs, thus resulting in higher treatment efficacy. In recent years, therapeutic drug monitoring (TDM) of antipsychotic drugs has proven to be of great value for assessing poor adherence in patients and determining genetic variability in antipsychotic metabolism.[125] Pharmacotherapy treatment of schizophrenia is often carried out with the simultaneous use of two or more antipsychotic drugs to achieve sufficient control of psychotic systems. Therefore reliable, sensitive, and selective analytical methods are required and have a very important role to play in adherence studies of antipsychotic drugs.

DEPRESSION

Depression is one of the most prevalent forms of chronic mental disorders, with a huge public health impact globally. The treatment of depression requires prolonged pharmacotherapy with antidepressants.

Table 4.7 Bioanalytical Assays to Assess Adherence to Antidepressant Drugs

Analytical Method	Biosample	Analyte(s)	Location	References
HPLC	Plasma	Antidepressants—venlafaxine, fluoxetine, paroxetine, citalopram	United States	126
HPTLC	Serum	Antidepressants	Chile	127–129
HPLC, LC-MS/MS	Plasma	Antidepressants—fluoxetine, paroxetine, sertraline, venlafaxine (HPLC) Citalopram, escitalopram, clomipramine (LC-MS/MS)	The Netherlands	130
LC-MS/MS (qqq)	Urine	Antidepressants	United States	131

Nonadherence to antidepressant medications is a significant barrier to the successful treatment of depression in clinical practice[132] and the World Health Organization has recognized depression as one of the chronic conditions to be focused on for improving adherence to medications.[1] There are two main aspects of adherence problems with antidepressants—the early discontinuation of antidepressant pharmacotherapy and the lack of consistency with the prescribed regimen in the context of ongoing use. The risk of relapse and additional depressive episodes increases with early discontinuation of an antidepressant medicine. Medication nonadherence in the context of ongoing use is also a common problem with antidepressants. For example, patients may either intentionally or unintentionally miss doses, take extra doses, delay the timing of their doses, or take drug holidays. This lack of medication adherence could also affect treatment outcomes resulting in nonresponse or discontinuation symptoms.[132] A number of bioanalytical assays used to investigate adherence to antidepressants are summarized in Table 4.7 but this data excludes work carried out by private health laboratories.[112,113]

Objective measures to assess adherence to antidepressants have been conducted using bioanalytical assays for plasma, serum, and urine samples (Table 4.7). Several HPLC and LC-MS/MS analytical methodologies have been reported for the identification and quantification of antidepressants in these different biological matrices.

BIPOLAR DISORDER

Bipolar disorder is a chronic psychiatric condition that causes unusual shifts in mood, energy, activity levels, and the ability to carry out

Table 4.8 Bioanalytical Assays to Assess Adherence to Lithium				
Analytical Method	Biosample	Analyte(s)	Location	References
Flame photometry	Whole blood, plasma	Lithium carbonate	Brazil	133
Ion selective electrodes, flame photometry	Serum	Lithium carbonate	Denmark	134
Ion selective electrodes	Plasma	Lithium carbonate	India	135

day-to-day tasks. Symptoms of bipolar disorder are severe and can result in damaged relationships, poor job or school performance, and even suicide. But bipolar disorder can be treated, and people with this illness can lead full and productive lives.[136] Lithium is the main treatment for bipolar disorder[137] but estimated medication adherence is low and ranges from 20% to 64%.[138] Although effective, lithium requires specific care as it has a narrow therapeutic range, high risk of intoxication, and slow onset of action. Furthermore, lithium serum levels vary as a result of different pharmacological interactions, and elimination and plasma level variations among patients have been reported.[139] The TDM of lithium is therefore very important for the effectiveness of the drug therapy and patient safety, and in Belgium a patient self-test at home kit to determine lithium levels is available from Medivate BV.[140] In this test a single drop of blood from a fingerprick is placed on a lab-chip which is then placed in a reader and after 9 minutes the lithium level is displayed. In the National Health Service in the United Kingdom the checking of blood lithium levels is specified by the Area Prescribing Committee and includes weekly monitoring 7 days after the initiation of dose until stable and then every 3 months unless there is a medication change.[141] A different range of analytical methods are recorded in Table 4.8 which have been necessary to assess adherence to lithium pharmacotherapy.

Objective measurements of adherence to prescribed lithium have been carried out on blood samples using flame photometry and ion selective electrodes.

EPILEPSY

Epilepsy is a chronic neurological condition that affects people of all ages. Approximately 50 million people worldwide have epilepsy[142] which is associated with physical, psychological, and social problems. Patients with epilepsy have a higher rate of suicide, anxiety, depression, sudden unexplained death, and accidental death. Prolonged seizures may cause physical injury, neuronal death leading to cognitive impairment, and can be fatal.[143] Epilepsy continues to be a highly stigmatized and disabling condition,[144] requiring a lifelong process of adherence to the prescriber's instructions and drug therapy. Rates of medication adherence in epilepsy are estimated to range widely from 20% to 80%[1] and this reduces the benefit that could be gained from the prescribed medication. Poor adherence to antiepileptic drugs (AEDs) may be the most important cause of poorly controlled epilepsy resulting in seizure relapses. In epilepsy, adherence to medication is commonly assessed by the direct method of measuring biological specimen AED levels.[145] Various objective studies have investigated the adherence to prescribed AEDs using analysis of biological samples (Table 4.9).

Direct measures of adherence to AEDs have been conducted by the measurement of drug levels in blood plasma or serum, DBS, hair, and saliva. In a resource-limited setting in rural Kenya,[146] immunoassay

Table 4.9 Bioanalytical Assays to Assess Adherence to Antiepileptic Drugs				
Analytical Method	Biosample	Analyte(s)	Location	References
HPLC	Hair	Antiepileptic drug—carbamazepine, lamotrigine	United Kingdom	147
	Hair, plasma	Antiepileptic drugs—carbamazepine	United Kingdom	148
	Saliva, plasma	Antiepileptic drug—lamotrigine	Australia	149
	DBS	Antiepileptic drugs—lamotrigine, oxcarbazepine	The Netherlands	150
HPLC, GC	DBS	Antiepileptic drugs	United Kingdom	151
	Plasma	Antiepileptic drugs—carbamazepine, lamotrigine, levetiracetam, oxcarbazepine, phenobarbital, topiramate, valproic acid	France	152
HPLC, GC-MS	DBS, serum	Antiepileptic drugs—valproic acid, carbamazepine, phenobarbital, phenytoin	Brazil	153
HPTLC	Serum	Antiepileptic drug—lamotrigine	Chile	154
Immunoassay	Plasma	Antiepileptic drugs—carbamazepine, lamotrigine, levetiracetam, oxcarbazepine, phenobarbital, topiramate, valproic acid	France	152
	Blood	Antiepileptic drugs—carbamazepine, phenobarbital, phenytoin	Kenya	146
	Saliva, plasma	Antiepileptic drug—topiramate	United States	155

has been employed to analyze blood samples. Typically AED medication levels have been measured in such biological samples by conducting HPLC-, HPTLC-, or GC-based assays.

Plasma and serum monitoring remains the standard for judging adherence to epilepsy medication regimens and sampling intervals for these biological samples can be as long as several months. A drop in medication level of a certain predetermined percentage, dependent on study and medication being studied, indicates nonadherence.[145] The use of minimally invasive DBS sampling, which can be performed by nonexperts, has shown great potential for assessing adherence to AEDs in children via home sampling.[151] The key findings from these investigations were that 33 of the 100 children studied were nonadherent to their AED medication and that the DBS mode of sample collection could effectively be used in adherence assessment. Self-collection of DBS samples by a patient has also been used in a study to investigate changes in pharmacokinetics of AEDs during pregnancy.[150] In this study the blood concentration levels from DBS samples were used to inform the doses prescribed for two patients (A and B) taking a combination of lamotrigine (LTG) and oxcarbazepine (OXC) during pregnancy. Baseline levels for both drugs were determined from DBS samples to give the blood serum reference ranges prior to pregnancy. To maintain serum levels within reference ranges during pregnancy the AED dose levels were changed significantly for patient A where the OXC daily dose was increased from 1800 to 2400 mg and the LTG daily dose was increased from 275 to 675 mg. For patient B the OXC dose was increased from 900 to 2100 mg daily and the daily LTG dose changed from 600 to 1000 mg. Both of these case studies clearly indicate the need for personalization of medicines. A study to evaluate adherence to antiepileptic pharmacotherapy compared a standardized questionnaire with measured AED concentrations in serum and in DBS. The results showed that AED concentrations in DBS were proportional to those obtained in serum samples and that adherence to the treatment with anticonvulsants was 66.6% using the standardized questionnaire and 63.3% through the laboratory analyses.[153]

Adherence studies using the determination of AED levels in human hair have shown that this mode of sampling is easy to perform and has similar sensitivity to blood plasma.[148] Other studies consider hair sampling more accurate than self-report methods or blood plasma monitoring.[147]

Sampling saliva for AED levels to assess medication adherence offers numerous advantages. It is not painful to collect a sample which is an important consideration for pediatric patients. Furthermore, it does not depend on venous access which is therefore a benefit for pediatric or geriatric patients who may have few veins.[145] It is reported that patients preferred saliva samples over blood sampling but saliva samples may be more readily contaminated[156] and it is necessary to measure the individual rate of saliva production on a daily basis.[149]

MALARIA

Malaria is a serious public health concern with approximately 3.2 billion people—almost half of the world's population—at risk. This communicable disease continues to cause more than 200 million cases and more than 400,000 deaths per year, of which 91% occurred in resource-limited Sub-Saharan Africa. Additionally, 70% of all malaria deaths occur in children under 5 years of age.[157] Malaria is an acute mosquito-borne infection caused by the *Plasmodium falciparum* parasite. The disease is preventable and treatable with antimalarial drugs but does require immediate treatment after the onset of symptoms. A high level of adherence to malaria treatment is important in ensuring drug effectiveness and contributes to the patient's complete recovery and prevents the emergence of parasite resistance to the antimalarial drug. The assessment of adherence among malaria patients faces many of the same problems presented for the HIV/AIDS situation, and in parallel with HIV/AIDS, multimethod assessment processes have been used to investigate adherence to prescription. These include pill counts and patient or carer interviews in conjunction with the direct methods summarized in Table 4.10. Three different, but complementary, direct methods were used:

- Quantitative PCR (qPCR) analysis of a biological specimen extract was used, as described in Chapter 3, to identify the malaria parasites and to determine adherence to prescription based on the number of parasites present
- Microscope analysis of blood smears to identify the malaria parasites and to determine adherence to prescription based on the number of parasites present
- HPLC-UV assay of a patient plasma extract to quantify the level of antimalarial drug(s) in the biological sample

Table 4.10 Bioanalytical Assays to Assess Adherence to Antimalarial Drugs				
Analytical Method	Biosample	Analyte(s)	Location	References
qPCR	Plasma	*Plasmodium falciparum*	Bangladesh → Thailand	158
			Tanzania → Sweden	159
	DBS	*Plasmodium falciparum*	Ghana, Nigeria, Uganda	160
Microscopy	Blood smear	*Plasmodium falciparum*	Uganda	161
			Tanzania → Sweden	159
			Ghana, Nigeria, Uganda	160
		Parasitemia	Nigeria, Tanzania	162,163
HPLC	Plasma	Sulfadoxine	Uganda	161
		Dapsone, lumefantrine	United Kingdom → Malawi	164
		Primaquine	Thailand	165
		Lumefantrine	Nigeria, Tanzania	162,163
	DBS	Sulphadoxine, pyrimethamine, chloroquine	Senegal → France	166
		Lumefantrine	Tanzania → Sweden	159
			Tanzania	167
LC-QTRAP	DBS	Primaquine	Thailand	168

Access to these sophisticated analytical methods meant some biosamples were transferred between countries as indicated by →.

A high degree of international collaboration in sample collection and analysis is evident from Table 4.10 and again the need to transport samples safely for extended time periods is demonstrated. The entry United Kingdom → Malawi indicates that a team was sent from the United Kingdom to carry out sample collection and analysis on-site in Malawi relying on extensive sample pre-work-up to facilitate successful HPLC analysis.[164]

There is clearly a need for medication adherence monitoring during treatment regimens for antimalarials in personalized pharmacotherapy. This necessitates calculation of the levels of these drugs in biological samples. In the low resource settings where malaria is highly prevalent, concentrations of antimalarial drugs have predominantly been measured in plasma, serum, and capillary blood sampled as DBS using HPLC. Often these measurements have been coupled with PCR analysis and/or microscopy to assess the level of parasite. This requires robust, sensitive, specific, and reproducible bioanalytical methods. The robustness and cost-effectiveness of the DBS sampling approach have supported the evaluation of filed samples in Tanzania for malaria patients treated with artemether and lumefantrine.[159,167] Using HPLC analysis, blood lumefantrine levels could be estimated accurately to assess adherence to treatment up to 4 days after the last dose of a standard regimen. The median lumefantrine concentration was statistically significantly lower in patients with recrudescence (97 ng/mL), compared with reinfections (205 ng/mL), or no parasites reappearance (217 ng/mL) and the PCR-assessed cure rate by day 42 of treatment was 93%.[159] Elsewhere it has been noted that these adherence levels are only maintained where the patients are involved in specific tests.[169] In another study the effectiveness of the home management of malaria for children was assessed across selected sites from Ghana, Nigeria, and Uganda by comparing thick blood smears and the PCR analysis of DBS samples with carer reports of adherence to a prescribed artemisinin.[160] In this study, 1740 children were initially recruited with 1189 exhibiting parasitemia at baseline. After a 28-day regimen of home treatment only 51% of parasitemic children reported for the follow-up test. The use of DBS in the bioanalysis of antimalarials has been recognized, given the value of the reduction in sample volume it offers and its potential for home sampling. A problem reported with this sampling format is the quantification of artemisinin derivatives.[170] These compounds decompose in the presence of iron (II) and are therefore unstable for long periods in a blood spot.

IMMUNOSUPPRESSION

Following solid organ transplantation, lifelong adherence to immunosuppressant medication regimens is a vital necessity to prevent transplant rejection, graft loss, and additional morbidity. Despite this, nonadherence to immunosuppressant treatment is highly prevalent for transplant patients, especially among pediatric transplant recipients, with a broad adherence rate range from 25% to 100% reported.[171,172] Immunosuppressant drugs have narrow therapeutic windows and highly variable pharmacokinetics and adherence monitoring is therefore crucial to provide individualized pharmacotherapy and to avoid toxicity-related overdosage and transplant rejection from underdosage.[173] This requires frequent hospital visits to phlebotomists. The direct measurement of immunosuppressant drugs for TDM purposes has been mainly based on immunoassays (Table 4.11) which are suited for

Table 4.11 Bioanalytical Assays to Assess Adherence to Immunosuppression Drugs				
Analytical Method	Biosample	Analyte(s)	Location	References
LC-MS/MS (qqq)	Blood	Tacrolimus, sirolimus, everolimus	Australia	174
	DBS	Creatinine, tacrolimus, sirolimus, everolimus, cyclosporine A	The Netherlands	175
		Tacrolimus, creatinine	United States	176
		Tacrolimus, cyclosporin A	United Kingdom	177
Immunoassay	Blood	Cyclosporine, rifampin, tacrolimus, everolimus	Spain	178
		Tacrolimus, mycophenolic acid	United States	179
	Plasma	Tacrolimus	United States	180–182
		Tacrolimus, everolimus, azathioprine, prednisone, mycophenolate mofetil, cyclosporine, sirolimus	France	183

routine laboratory analyses with excellent automation and high throughput. As reviewed by Taylor and coworkers,[173] a lack of specificity for the parent drug and variable overestimation of the immunosuppressant drugs tacrolimus and cyclosporine are major drawbacks of this analytical method. A shift to HPLC coupled with tandem mass spectrometry (MS/MS) has been apparent in recent times due to the superior selectivity and specificity offered by this analytical method. These advances in mass spectrometry techniques have facilitated capillary blood sampling onto DBSs and thus enable adherence monitoring of transplant recipients to be carried out at home.

ASTHMA

Asthma is a major noncommunicable chronic disease of the air passages of the lungs which inflames and narrows them. It is estimated that as many as 235 million people of all ages suffer from asthma[184] and the burden of this disease to healthcare systems, carers, and patients is escalating worldwide. Poor adherence to asthma medication regimens has been reported in both children and adults with adherence rates ranging from 30% to 70% documented.[185,186] Direct measures have been used to assess adherence with asthma medications as summarized in Table 4.12.

Table 4.12 Bioanalytical Assays to Assess Adherence to Asthma Drugs

Analytical Method	Biosample	Analyte(s)	Location	References
LC-MS/MS (qqq)	Urine	Fluticasone propionate	United States	187,188
	Urine, blood, sputum	Beclomethasone, betamethasone, budesonide, dexamethasone, flunisolide, fluticasone propionate, methylprednisolone, prednisolone, predinisone	United States	189
Immunoassay	DBS	Theophylline	United Kingdom	190

Direct measures, which confirm the use of the asthma medication by assaying it in blood, urine, and sputum using LC-MS/MS have been documented. However, in a very forward-thinking study in 1988[190] the analysis of DBS samples by immunoassay was used to assess levels of theophylline in the blood of 100 children receiving treatment for asthma. One of the purposes of this study was to assess the willingness of patients from 14 months to 15 years old to provide DBS samples collected at home. Of the 62 samples returned, 37 showed less than the therapeutic concentration of the target drug, 3 that were at potentially toxic levels, and 6 children had levels less than the detection limits, suggesting they were nonadherent to the prescribed asthma medication.

OTHER INFECTIONS

Antimicrobial resistance results from inappropriate use of antibiotics and makes common or life-threatening infections more difficult or sometimes impossible to treat. Proper adherence to antibiotic pharmacotherapy is crucial to prevent antimicrobial resistance and to maintain therapeutic drug concentration in blood and for achieving therapeutic benefits. Despite this, a significant number of patients are not adherent with basic medication for infectious diseases, including antibiotics, which are one of the most commonly prescribed medicines worldwide.[191] Tuberculosis (TB) is an infectious disease which is a major killer worldwide and the largest number of reported cases in 2014 were in India, Indonesia, Nigeria, Pakistan, China, and South Africa. It is caused by bacteria (*Myobacterium tuberculosis*) that most often affect the lungs and is spread from person to person through the air.[192] This

Table 4.13 Bioanalytical Assays to Assess Adherence to Antibiotics, Antifungal Drugs, and Antituberculosis Drugs				
Analytical Method	Biosample	Analyte(s)	Location	References
PCR	Urine	*Chlamydia tracomatis*	United States	193
HPLC	Urine	Antibiotic—trimethorprim	United States	194
	DBS	Antibiotic—ciprofloxacin	Jordan	195
	Saliva	Antifungal—clotrimazole	Israel	196
LC-MS	DBS	Antibiotic—rifaximin	India	197
LC-MS/MS (qqq)	Urine	Antibiotic—mupirocin	United Kingdom	198
		Tuberculosis drug—isoniazid	South Africa	199
	Serum	Tuberculosis drugs—moxifloxacin, prothionamide, cycloserine	Korea	200

is a curable disease if drug treatment is received quickly and appropriately. Nonadherence to anti-TB drug therapy contributes to the development of multidrug-resistant TB (MDR-TB), prolonged infectiousness, and poor TB treatment outcomes.[201,202]

Direct measures to assess adherence to antibiotic and antifungal drugs have been conducted using bioanalytical assays for urine, saliva, and DBS samples (Table 4.13). The earlier assays used PCR while more recent assays have used mass spectrometry−based methods.

OTHER CLINICAL APPLICATIONS

Adherence studies have also been carried out in areas relating to attention deficit hyperactivity disorder,[203] inflammatory bowel disease,[204,205] rheumatoid arthritis,[206] and gout[207] principally based on analyses of serum or plasma by LC-MS/MS techniques. This work has been carried out in the United Kingdom and United States.

CONCLUSIONS

In reality, it is not possible to conclude exactly how much work has been done on the direct measurement of adherence to medication because details of some of this work are restricted by hospital/patient confidentiality and the confidentiality maintained by private analytical laboratories. Suffice it to say that the amount and extent of the work reported here demonstrates the recognition of the need to carry out medication adherence monitoring and thus provides an objective measure to personalize medication and that "one size fits all" is no longer appropriate. Indeed pharmacogenetic tests followed by optimizing dosing based on direct measurement to assist TDM and pharmacokinetics appear to be the accepted approach worldwide. Quick, reliable, and affordable analytical methods should be used more in the routine assessment of patient biosamples for medication adherence. In a recent funding application, a representative from the British Heart Foundation commented that this approach had great potential to make a step change in allowing clinician–patient discussions to be focused around which treatments are being taken, which are not and for what reasons.

REFERENCES

1. Sabaté E. *Adherence to Long-Term Therapies: Evidence for Action.* Geneva, Switzerland: World Health Organization; 2003.

2. *British Heart Foundation Centre on Population Approaches for Non-communicable Disease Prevention.* Cardiovascular disease statistics 2014. < https://www.bhf.org.uk/ ~ /media/files/ research/heart-statistics/bhf_cvd-statistics-2014_web.pdf > ; 2014 Accessed 26.11.15.

3. González O, Ferreirós N, Blanco ME, Alonso RM. Cardiovascular drug determination in bioanalysis: an update. *Bioanalysis.* 2015;7(18):2399–2417. Available from: http://dx.doi.org/ 10.4155/bio.15.163.

4. Dawson J, Quinn T, Rafferty M, et al. Aspirin resistance and compliance with therapy. *Cardiovasc Ther.* 2011;29(5):301–307. Available from: http://dx.doi.org/10.1111/j.1755-5922.2010.00188.x.

5. Mennickent S, Vega MM, Vega HM, De Diego M, Fierro R. Quantitative determination of propranolol in human serum by high-performance thin-layer chromatography. *JPC-J Planar Chromatogr.* 2012;25(1):54–59.

6. Mennickent S, Contreras J, Schulz B, de Diego M, Vega M. High performance thin layer chromatographic determination of nifedipine in human serum after liquid-liquid extraction. *Quim Nova.* 2012;35(2):411–415.

7. Schmieder RE, Ott C, Veelken R, et al. 7B.01: final analysis on adherence to antihypertensive medication in treatment resistant hypertension (TRH) undergoing (TRH) renal denervation (RDN). *J Hypertens.* 2015;33((suppl 1)):e92. Available from: http://dx.doi.org/10.1097/01. hjh.0000467598.27100.ca.

8. Ceral J, Habrdova V, Vorisek V, et al. Difficult-to-control arterial hypertension or uncooperative patients? The assessment of serum antihypertensive drug levels to differentiate non-responsiveness from non-adherence to recommended therapy. *Hypertens Res.* 2011;34 (1):87–90. Available from: http://dx.doi.org/10.1038/hr.2010.183.

9. Lawson AJ, Shipman KE, George S, Dasgupta I. A novel 'dilute-and-shoot' liquid chromatography-tandem mass spectrometry method for the screening of antihypertensive drugs in urine. *J Anal Toxicol.* 2016;40(1):17–27.

10. Hamdidouche I, Jullien V, Billaud EM, et al. 7B.06: routine urinary detection of anti-hypertensive drugs for estimation of adherence to treatment: a cross sectional study. *J Hypertens.* 2015;33(suppl 1):e93. Available from: http://dx.doi.org/10.1097/01.hjh. 0000467603.33785.69.

11. Patel P, Gupta PKC, White CMJ, et al. Screening for non-adherence to antihypertensive treatment as a part of a diagnostic pathway to renal denervation. *J Hum Hypertens.* 2015;:1−6. Available from: http://dx.doi.org/10.1038/jhh.2015.103.

12. Tomaszewski M, White C, Patel P, et al. High rates of non-adherence to antihypertensive treatment revealed by high-performance liquid chromatography-tandem mass spectrometry (HP LC-MS/MS) urine analysis. *Heart.* 2014;100:855−861. Available from: http://dx.doi.org/10.1136/heartjnl-2013-305063.

13. Florczak E, Tokarczyk B, Warchol-Celinska E, et al. Assessment of adherence to treatment in patients with resistant hypertension using toxicological serum analysis. *Pol Arch Med Wewn.* 2015;125(1−2):65−72.

14. Strauch B, Petrak O, Zelinka T, et al. Precise assessment of noncompliance with the antihypertensive therapy in patients with resistant hypertension using toxicological serum analysis. *J Hypertens.* 2013;31(12):2455−2461.

15. Dias E, Hachey B, McNaughton C, et al. An LC-MS assay for the screening of cardiovascular medications in human samples. *J Chromatogr B Analyt Technol Biomed Life Sci.* 2013;937:44−53. Available from: http://dx.doi.org/10.1016/j.jchromb.2013.08.010.

16. Zoerner AA, Schroeder C, Kayacelebi AA, et al. A validated, rapid UPLC-MS/MS method for simultaneous ivabradine, reboxetine, and metoprolol analysis in human plasma and its application to clinical trial samples. *J Chromatogr B Analyt Technol Biomed Life Sci.* 2013;927:105−111. Available from: http://dx.doi.org/10.1016/j.jchromb.2013.01.016.

17. Jung O, Gechter JL, Wunder C, et al. Resistant hypertension? Assessment of adherence to toxicological urine analysis. *J Hypertens.* 2013;31(4):766−774.

18. Lawson G, Cocks E, Tanna S. Quantitative determination of atenolol in dried blood spot samples by LC-HRMS: a potential method for assessing medication adherence. *J Chromatogr B Analyt Technol Biomed Life Sci.* 2012;897:72−79. Available from: http://dx.doi.org/10.1016/j.jchromb.2012.04.013.

19. Lawson G, Cocks E, Tanna S. Bisoprolol, ramipril and simvastatin determination in dried blood spot samples using LC-HRMS for assessing medication adherence. *J Pharm Biomed Anal.* 2013;81−82:99−107. Available from: http://dx.doi.org/10.1016/j.jpba.2013.04.002.

20. Tanna S, Bernieh D, Lawson G. LC-HRMS analysis of dried blood spot samples for assessing adherence to cardiovascular medications. *J Bioanal Biomed.* 2015;7(1):001−005. Available from: http://dx.doi.org/10.4172/1948-593X.1000115.

21. Helfer AG, Michely JA, Weber AA, Meyer MR, Maurer HH. Orbitrap technology for comprehensive metabolite-based liquid chromatographic-high resolution tandem mass spectrometric urine drug screening—exemplified for cardiovascular drugs. *Anal Chim Acta.* 2015;891:221−233. Available from: http://dx.doi.org/10.1016/j.aca.2015.08.018.

22. Ewen S, Meyer MR, Cremers B, et al. Blood pressure reductions following catheter-based renal denervation are not related to improvements in adherence to antihypertensive drugs measured by urine/plasma toxicological analysis. *Clin Res Cardiol.* August 26 2015; [Epub ahead of print].

23. Zantek ND, Luepker RV, Duval S, Miller K, Oldenburg N, Hirsch AT. Confirmation of reported aspirin use in community studies: utility of serum thromboxane B2 measurement. *Clin Appl Thromb Hemost.* 2014;20(4):385−392. Available from: http://dx.doi.org/10.1177/1076029613486537.

24. Azizi M, Ménard J, Peyrard S, Lièvre M, Marre M, Chatellier G. Assessment of patients' and physicians' compliance to an ACE inhibitor treatment based on urinary N-acetyl

Ser-Asp-Lys-Pro determination in the Noninsulin-Dependent Diabetes, Hypertension, Microalbuminuria, Proteinuria, Cardiovascular Events, and Ramipril (DIABHYCAR) study. *Diabetes Care.* 2006;29(6):1331−1336.

25. Vrijens B, Vincze G, Kristanto P, et al. Adherence to prescribed antihypertensive drug treatments: longitudinal study of electronically complied dosing histories. *BMJ.* 2008;336: 1114−1117.

26. Clark M. Adherence to treatment in patients with type 2 diabetes. *J Diabetes Nurs.* 2004;8 (10):386−391.

27. AbuRuz S, Millership J, McElnay J. Dried blood spot liquid chromatography assay for therapeutic drug monitoring of metformin. *J Chromatogr B Analyt Technol Biomed Life Sci.* 2006;832(2):202−207.

28. AbuRuz S, Millership J, McElnay J. Determination of metformin in plasma using a new ion pair solid phase extraction technique and ion pair liquid chromatography. *J Chromatogr B Analyt Technol Biomed Life Sci.* 2003;798(2):203−209.

29. Tipton JM. Overview of the challenges related to oral agents for cancer and their impact on adherence. *Clin J Oncol Nurs.* 2015;19(3 suppl):37−40. Available from: http://dx.doi.org/10.1188/15.S1CJON.37-40.

30. Barillet M, Prevost V, Joly F, Clarisse B. Oral antineoplastic agents: how do we care about adherence? *Br J Clin Pharmacol.* 2015;80(6):1289−1302. Available from: http://dx.doi.org/10.1111/bcp.12734.

31. Mathes T, Antoine SL, Pieper D, Eikermann M. Adherence enhancing interventions for oral anticancer agents: a systematic review. *Cancer Treat Rev.* 2014;40(1):102−108. Available from: http://dx.doi.org/10.1016/j.ctrv.2013.07.004.

32. McCue DA, Lohr LK, Pick AM. Improving adherence to oral cancer therapy in clinical practice. *Pharmacotherapy.* 2014;34(5):481−494.

33. D'Amato S. Improving patient adherence with oral chemotherapy. *Oncol Issue.* 2008;42−45: July/August Issue.

34. Marin D, Bazeos A, Mahon F-X, et al. Adherence is a critical factor for achieving molecular responses in patients with chronic myeloid leukemia who achieve complete cytogenetic responses on imatinib. *J Clin Oncol.* 2010;28(14):2381−2388.

35. Ishikawa Y, Kiyoi H, Watanabe K, et al. Trough plasma concentration of imatinib reflects BCR-ABL kinase inhibitory activity and clinical response in chronic-phase chronic myeloid leukemia: a report from the BINGO study. *Cancer Sci.* 2010;101(10):2186−2192. Available from: http://dx.doi.org/10.1111/j.1349-7006.2010.01643.x.

36. Yoshida C, Komeno T, Hori M, et al. Adherence to the standard dose of imatinib, rather than dose adjustment based on its plasma concentration, is critical to achieve a deep molecular response in patients with chronic myeloid leukemia. *Int J Hematol.* 2011;93(5):618−623. Available from: http://dx.doi.org/10.1007/s12185-011-0838-3.

37. Sharma SK, Kumar S, Vijayakumar AR, et al. Utility of the trough plasma imatinib level monitoring at two time points in patients with the chronic myeloid leukemia-chronic phase. *J Cancer Res Ther.* 2014;10(2):305−308. Available from: http://dx.doi.org/10.4103/0973-1482.136583.

38. Režonja Kukec R, Grabnar I, Mrhar A, et al. A simple dried blood spot method for clinical pharmacological analyses of etoposidein cancer patients using liquid chromatography and fluorescence detection. *Clin Chim Acta.* 2016;452:99−105. Available from: http://dx.doi.org/10.1016/j.cca.2015.10.026.

39. Clarke Hillyer G, Neugut AI, Crew KD, et al. Use of a urine anastrozole assay to determine treatment discontinuation among women with hormone-sensitive breast cancer: a pilot study. *J Oncol Pract.* 2012;8(5):e100−e104. Available from: http://dx.doi.org/10.1200/JOP.2011.000487.

40. Timmers L, Boons CC, Mangnus D, et al. The use of erlotinib in daily practice: a study on adherence and patients' experiences. *BMC Cancer*. 2011;11:284. Available from: http://dx. doi.org/10.1186/1471-2407-11-284.

41. Beer B, Schubert B, Oberguggenberger A, et al. Development and validation of a liquid chromatography-tandem mass spectrometry method for the simultaneous quantification of tamoxifen, anastrozole, and letrozole in human plasma and its application to a clinical study. *Anal Bioanal Chem*. 2010;398(4):1791–1800. Available from: http://dx.doi.org/10.1007/s00216-010-4075-z.

42. Oberguggenberger AS, Sztankay M, Beer B, et al. Adherence evaluation of endocrine treatment in breast cancer: methodological aspects. *BMC Cancer*. 2012;12:474. Available from: http://dx.doi.org/10.1186/1471-2407-12-474.

43. Antunes MV, Raymundo S, de Oliveira V, et al. Ultra-high performance liquid chromatography tandem mass spectrometric method for the determination of tamoxifen, *N*-desmethyl tamoxifen, 4-hydroxytamoxifen and endoxifen in dried blood spots—development, validation and clinical application during breast cancer adjuvant therapy. *Talanta*. 2015;132:775–784. Available from: http://dx.doi.org/10.1016/j.talanta.2014.10.040.

44. Brier MJ, Chambless D, Gross R, et al. Association between self-report adherence measures and oestrogen suppression among breast cancer survivors on aromatase inhibitors. *Eur J Cancer*. 2015;51(14):1890–1896. Available from: http://dx.doi.org/10.1016/j.ejca.2015.06.113.

45. Pai AL, McGrady ME. Assessing medication adherence as a standard of care in pediatric oncology. *Pediatr Blood Cancer*. 2015;62(suppl 5)):S696–S706. Available from: http://dx.doi. org/10.1002/pbc.25795.

46. Spoelstra SL. Issues related to overadherence to oral chemotherapy or targeted agents. *Clin J Oncol Nurs*. 2013;17(6):604–609. Available from: http://dx.doi.org/10.1188/13.CJON.17-06AP.

47. Tanna S, Lawson G. Dried blood spot analysis to assess medication adherence and to inform personalization of treatment. *Bioanalysis*. 2014;6(21):2825–2838.

48. *World Health Organisation—Media Centre*. HIV/AIDS fact sheet no 360. < http://www.who. int/mediacentre/factsheets/fs360/en/ > ; 2015 Accessed 30.11.15.

49. Gordon LL, Gharibian D, Chong K, Chun H. Comparison of HIV virologic failure rates between patients with variable adherence to three antiretroviral regimen types. *AIDS Patient Care STDS*. 2015;29(7):384–388. Available from: http://dx.doi.org/10.1089/apc.2014.0165.

50. Spire B, Carrieri P, Sopha P, et al. Adherence to antiretroviral therapy in patients enrolled in a comprehensive care program in Cambodia: a 24-month follow-up assessment. *Antivir Ther*. 2008;13(5):697–703.

51. Garrido C, Zahonero N, Corral A, et al. Correlation between human immunodeficiency virus type 1 (HIV-1) RNA measurements obtained with dried blood spots and those obtained with plasma by use of Nuclisens EasyQ HIV-1 and Abbott Real Time HIV load tests. *J Clin Microbiol*. 2009;47(4):1031–1036. Available from: http://dx.doi.org/10.1128/JCM.02099-08.

52. Neogi U, Gupta S, Rodridges R, et al. Dried blood spot HIV-1 RNA quantification: a useful tool for viral load monitoring among HIV-infected individuals in India. *Indian J Med Res*. 2012;136(6):956–962.

53. Johannessen A, Garrido C, Zahonero N, et al. Dried blood spots perform well in viral load monitoring of patients who receive antiretroviral treatment in rural Tanzania. *Clin Infect Dis*. 2009;49(6):976–981. Available from: http://dx.doi.org/10.1086/605502.

54. Maiers TJ, Gous N, Nduna M, et al. An investigation of fingerstick blood collection for point-of-care HIV-1 viral load monitoring in South Africa. *S Afr Med J*. 2015;105 (3):228–231. Available from: http://dx.doi.org/10.7196/SAMJ.7799.

55. George L, Muro EP, Ndaro A, et al. Nevirapine concentrations in saliva measured by thin layer chromatography and self-reported adherence in patients on antiretroviral

therapy at Kilmanjaro Christian Medical Centre, Tanzania. *Ther Drug Monit.* 2014;36 (3):366−370.

56. Gandhi M, Yang Q, Bacchetti P, Huang Y. Short communication: a low-cost method for analyzing nevirapine levels in hair as a marker of adherence in resource-limited settings. *AIDS Res Hum Retroviruses.* 2014;30(1):25−28. Available from: http://dx.doi.org/10.1089/AID.2013.0239.

57. Demeter LM, Mukherjee AL, DiFrancesco R, et al. The design and implementation of A5146, a prospective trail assessing the utility of therapeutic drug monitoring using an inhibitory quotient in antiretroviral-experienced HIV-infected patients. *HIV Clin Trials.* 2008;9 (1):61−72. Available from: http://dx.doi.org/10.1310/hct0901-61.

58. Nieuwkerk PT, Sprangers MAG, Burger DM, et al. Limited patient adherence to highly active antiretroviral therapy for HIV-I infection in an observational cohort study. *Arch Intern Med.* 2001;161(16):1962−1968.

59. Arroyo MJH, Figueroa SEC, Correa RS, et al. Impact of a pharmaceutical care program on clinical evolution and antiretroviral treatment adherence: a 5-year study. *Patient Prefer Adherence.* 2013;7:729−739. Available from: http://dx.doi.org/10.2147/PPA.S47519.

60. de Matt MMR, Huitema ADR, Mulder JW, et al. Subtherapeutic antiretroviral plasma concentrations in routine clinical outpatient HIV care. *Ther Drug Monit.* 2003;25(3): 367−373.

61. Hugen PW, Burger DM, de Graaff M, et al. Saliva as a specimen for monitoring compliance but not for predicting plasma concentrations in patients with HIV treated with indinavir. *Ther Drug Monit.* 2000;22(4):437−445.

62. Van Schooneveld T, Swindells S, Nelson SR, et al. Clinical evaluation of a dried blood spot assay for atazanavir. *Antimicrob Agents Chemother.* 2010;54(10):4124−4128. Available from: http://dx.doi.org/10.1128/AAC.00297-10.

63. Bulgiba A, Mohammed UY, Chik Z, Lee C, Peramalah D. How well does self-reported adherence fare compared to therapeutic drug monitoring in HAART? *Prev Med.* 2013;57: S34−S36. Available from: http://dx.doi.org/10.1016/j.ypmed.2013.01.002.

64. Kumar S, Rao PSS, Earla R, Kumar A. Drug-drug interactions between anti-retroviral therapies and drugs of abuse in HIV systems. *Expert Opin Drug Metab Toxicol.* 2015;11 (3):343−355.

65. Liu X, Ma Q, Zhang FJ. Therapeutic drug monitoring in highly active antiretroviral therapy. *Expert Opin Drug Saf.* 2010;9(5):743−758.

66. Rakhmanina NY, van den Anker JN, Soldin SJ, et al. Can therapeutic drug monitoring improve pharmacotherapy of HIV infection in adolescents? *Ther Drug Monit.* 2010;32 (3):272−281.

67. Jones G, Hawkins K, Mullin R, et al. Understanding how adherence goals promote adherence behaviours: a repeated measure observational study with HIV seropositive patients. *BMC Public Health.* 2012;12(587). Available from: http://dx.doi.org/10.1186/1471-2458-12-587.

68. Koal T, Burhenne H, Römling R, et al. Quantification of antiretroviral drugs in dried blood spot samples by means of liquid chromatography/tandem mass spectrometry. *Rapid Commun Mass Spectrom.* 2005;19(21):2995−3001.

69. ter Heine R, Rosing H, van Gorp EC, et al. Quantification of protease inhibitors and non-nucleoside reverse transcriptase inhibitors in dried blood spots by liquid chromatography-triple quadrupole mass spectrometry. *J Chromatogr B Analyt Technol Biomed Life Sci.* 2008;867(2):205−212. Available from: http://dx.doi.org/10.1016/j.jchromb.2008.04.003.

70. Baxi SM, Liu A, Bacchetti P, et al. Comparing the novel method of assessing PrEP adherence/exposure using hair samples to other pharmacologic and traditional measures. *J AIDS.* 2015;68(1):13−20.

71. Shah SAB, Mullin R, Jones G, et al. Simultaneous analysis of antiretroviral drugs abacavir and tenofovir in human hair by liquid chromatography-tandem mass spectrometry. *J Pharm Biomed Anal.* 2013;74:308–313. Available from: http://dx.doi.org/10.1016/j.jpba.2012.10.023.

72. Olds PK, Kiwanuka JP, Nansera D, et al. Assessment of HIV antiretroviral therapy adherence by measuring drug concentrations in hair among children in rural Uganda. *AIDS Care.* 2015;27(3):327–332. Available from: http://dx.doi.org/10.1080/09540121.2014.983452.

73. Hickey MD, Salmen CR, Tessler RA, et al. Antiretroviral concentrations in small hair samples as a feasible marker of adherence in rural Kenya. *J AIDS.* 2014;66(3):311–315.

74. Gandhi M, Ameli N, Bacchettie P, et al. Protease inhibitor levels in hair strongly predict virologic response to treatment. *AIDS.* 2009;23(4):471–478.

75. Gras A, Schneider S, Karasi J-C, et al. Evaluation of saliva as an alternative matrix for monitoring plasma zidovudine, lamivudine and nevirapine concentrations in Rwanda. *Curr HIV Res.* 2011;9:000-000.

76. *AIDS info.* HIV treatment: the basics. < https://aidsinfo.nih.gov/education-materials/factsheets/21/51/hiv-treatment--the-basics > ; 2015 Accessed 30.11.15.

77. Hinkin CH, Hardy DJ, Mason KI, et al. Medication adherence in HIV-infected adults: effect of patient age, cognitive status, and substance abuse. *AIDS (London, England).* 2004;18 (suppl 1):S19–S25.

78. Roberts T, Bygrave H, Fajardo E, Ford N. Challenges and opportunities for the implementation of virological testing in resource-limited settings. *J Int AIDS Soc.* 2012;15(2):17324. Available from: http://dx.doi.org/10.7448/IAS.15.2.17324.

79. Ford N, Darder M, Spelman T, et al. Early adherence to antiretroviral medication as a predictor of long-term HIV virological suppression: five-year follow up of an observational cohort. *PLoS One.* 2010;5(5):e10460. Available from: http://dx.doi.org/10.1371/journal.pone.0010460.

80. ter Heine R, Beijnen JH, Huitema AD. Bioanalytical issues in patient-friendly sampling methods for therapeutic drug monitoring: focus on antiretroviral drugs. *Bioanalysis.* 2009;1 (7):1329–1338. Available from: http://dx.doi.org/10.4155/bio.09.124.

81. Barbosa J, Faria J, Carvalho F, et al. Hair as an alternative matrix in bioanalysis. *Bioanalysis.* 2013;5(8):895–914. Available from: http://dx.doi.org/10.4155/bio.13.50.

82. *World Health Organisation.* Media Centre. Hepatitis C. < http://www.who.int/mediacentre/factsheets/fs164/en/ > ; 2015 Accessed 15.01.16.

83. Trombatt WD, Koerner PH, Craft ZN, Miller RT, Kamal KM. Retrospective analysis of the medication utilization and clinical outcomes of patients treated with various regimens for Hepatitis C infection. *J Pharm Pract.* 2016;:pii:0897190015626008. [Epub ahead of print].

84. *Public Health England.* Hepatitis C in the UK. 2015 Report. < https://www.gov.uk/government/uploads/system/uploads/attachment_data/file/448710/NEW_FINAL_HCV_2015_IN_THE_UK_REPORT_28072015_v2.pdf > ; 2015 Accessed 18.01.16.

85. Hickman M, McDonald T, Judd A, et al. Increasing the uptake of hepatitis C virus testing among injecting drug users in specialist drug treatment and prison settings by using dried blood spots for diagnostic testing: a cluster randomized controlled trial. *J Viral Hepat.* 2008;15(4):250–254.

86. Craine N, Parry J, O'Toole J, D'Arcy S, Lyons M. Improving blood-borne viral diagnosis; clinical audit of the uptake of dried blood spot testing offered by a substance misuse service. *J Viral Hepat.* 2009;16(3):219–222. Available from: http://dx.doi.org/10.1111/j.1365-2893.2008.01061.x.

87. Tait JM, Stephens BP, McIntyre PG, Evans M, Dillon JF. Dry blood spot testing for hepatitis C in people who injected drugs: reaching the populations other tests cannot reach. *Frontline Gastroenterol.* 2013;4(4):255–262.

88. McMillin GA, Slawson MH, Marin SJ, Johnson-Davis KL. Demystifying analytical approaches for urine drug testing to evaluate medication adherence in chronic pain management. *J Pain Palliat Care Pharmacother*. 2013;27(4):322−339. Available from: http://dx.doi.org/10.3109/15360288.2013.847889.

89. Jain R, Partanayak RD. Pattern of consecutive urinanalysis results for patients on take home buprenorphine-naloxonone in India: do patients take it at home? *J Subst Use*. 2014;17 (3):230−234.

90. Berndt S, Maier C, Schütz HW. Polymedication and medication compliance in patients with chronic non-malignant pain. *Pain*. 1993;52(3):331−339.

91. Wasan AD, Michna E, Janfaza D, et al. Interpreting urine drug tests: prevalence of morphine metabolism to hydromorphone in chronic pain patients treated with morphine. *Pain Med*. 2008;9(7):918−923. Available from: http://dx.doi.org/10.1111/j.1526-4637.2007.00354.x.

92. Reisfield GM, Salazar E, Bertholf RL. Rational use and interpretation of urine drug testing in chronic opioid therapy. *Ann Clin Lab Sci*. 2007;37(4):301−314.

93. Lee HK, Lewis LD, Tsongalis GJ, et al. Negative urine opioid screening caused by rifampin-mediated induction of oxycodone hepatic metabolism. *Clin Chim Acta*. 2006;367 (1−2):196−200.

94. Vert P, Hamon I, Hubert C, Legagneur M, Hascoet JM. Infants of drug-addicted mothers: pitfalls of replacement therapy. *Bull Acad Natl Med*. 2008;192(5):961−969.

95. McMillin GA, Marin SJ, Johnson-Davis KL, Lawlor BG, Strathmann FG. A hybrid approach to urine drug testing using high-resolution mass spectrometry and select immunoassays. *Am J Clin Pathol*. 2015;143(2):234−240. Available from: http://dx.doi.org/10.1309/AJCPMJKBC6VJA1SP.

96. Kipping K, Maier C, Bussemas HH, Schwarzer A. Medication compliance in patients with chronic pain. *Pain Physician*. 2014;17(1):81−94.

97. Liu JC, Ma JD, Morello CM, Atayee RS, Best BM. Naltrexone metabolism and concomitant drug concentrations in chronic pain patients. *J Anal Toxicol*. 2014;38(4):212−217. Available from: http://dx.doi.org/10.1093/jat/bku019.

98. Melanson SE, Tanasijevic MJ, Snyder ML, Darragh A, Quade C, Jarolim P. Significant cost savings achieved by in-sourcing urine drug testing for monitoring medication compliance in pain management. *Clin Chim Acta*. 2013;422:10−14. Available from: http://dx.doi.org/10.1016/j.cca.2013.03.010.

99. Pesce A, West C, West R, et al. Determination of medication cutoff values in a pain patient population. *J Opioid Manag*. 2011;7(2):117−122.

100. West R, Pesce A, West C, et al. Observations of medication compliance by measurement of urinary drug concentrations in a pain management population. *J Opioid Manag*. 2010;6 (4):253−257.

101. Cone EJ, Zichterman A, Heltsley R, et al. Urine testing for norcodeine, norhydrocodone, and noroxycodone facilitates interpretation and reduces false negatives. *Forensic Sci Int*. 2010;198(1−3):58−61. Available from: http://dx.doi.org/10.1016/j.forsciint.2009.12.005.

102. Melanson SE, Ptolemy AS, Wasan AD. Optimizing urine drug testing for monitoring medication compliance in pain management. *Pain Med*. 2013;14(12):1813−1820. Available from: http://dx.doi.org/10.1111/pme.12207.

103. Pesce A, Rosenthal M, West R, et al. An evaluation of the diagnostic accuracy of liquid chromatography-tandem mass spectrometry versus immunoassay drug testing in pain patients. *Pain Physician*. 2010;13(3):273−281.

104. West R, Pesce A, West C, et al. Comparison of clonazepam compliance by measurement of urinary concentration by immunoassay and LC-MS/MS in pain management population. *Pain Physician*. 2010;13(1):71−78.

105. Manchikanti L, Malla Y, Wargo BW, et al. Protocol for accuracy of point of care (POC) or in-office urine drug testing (immunoassay) in chronic pain patients: a prospective analysis of immunoassay and liquid chromatography tandem mass spectrometry (LC/MS/MS). *Pain Physician.* 2010;13(1):E1−E22.

106. Mikel C, Almazan P, West R, et al. LC-MS/MS extends the range of drug analysis in pain patients. *Ther Drug Monit.* 2009;31(6):746−748. Available from: http://dx.doi.org/10.1097/FTD.0b013e3181c11fba.

107. Heltsley R, Depriest A, Black DL, et al. Oral fluid drug testing of chronic pain patients. II. Comparison of paired oral fluid and urine specimens. *J Anal Toxicol.* 2012;36(2):75−80. Available from: http://dx.doi.org/10.1093/jat/bkr019.

108. Heltsley R, DePriest A, Black DL, et al. Oral fluid drug testing of chronic pain patients. I. Positive prevalence rates of licit and illicit drugs. *J Anal Toxicol.* 2011;35(8):529−540.

109. Christo PJ, Manchikanti L, Ruan X, et al. Urine drug testing in chronic pain. *Pain Physician.* 2011;14(2):123−143.

110. McCarberg BH. A critical assessment of opioid treatment adherence using urine drug testing in chronic pain management. *Postgrad Med.* 2011;123(6):124−131. Available from: http://dx.doi.org/10.3810/pgm.2011.11.2502.

111. Moore C. Drug testing and adherence monitoring in pain management: oral fluid testing. *J Opiod Manag.* 2015;11(1):69−75. Available from: http://dx.doi.org/10.5055/jom.2015.0254.

112. *American Association for Clinical Chemistry.* < https://labtestsonline.org/ >; 2015 Accessed 15.01.16.

113. *American Society of Clinical Laboratory Science.* Direct access testing. < http://www.ascls.org/position-papers/177-direct-access-testing/139-directaccess-testing >; 2014 Accessed 15.01.16.

114. Piper C. Fraud Magazine. 10 popular health care provider fraud schemes. < http://www.fraud-magazine.com/article.aspx?id = 4294976280 >; 2013 Accessed 15.01.16.

115. Wu AHB, Lynch KL. Therapeutic drug monitoring to support clinical pharmacogenomics. In: Xu QA, Madden TL, eds. *LC-MS in Drug Bioanalysis.* New York, NY: Springer; 2012:127−141.

116. Haddad PM, Brain C, Scott J. Nonadherence with antipsychotic medication in schizophrenia: challenges and management strategies. *Patient Relat Outcome Meas.* 2014;5:43−62. Available from: http://dx.doi.org/10.2147/PROM.S42735.

117. Barkhof E, Meijer CJ, de Sonneville LMJ, Linszen DH, de Haan L. Interventions to improve adherence to antipsychotic medication in patients with schizophrenia—a review of the past decade. *Eur Psychiatry.* 2012;27:9−18.

118. Acosta FJ, Hernández JL, Pereira J, Herrera J, Rodríguez CJ. Medication adherence in schizophrenia. *World J Psychiatry.* 2012;2(5):74−82. Available from: http://dx.doi.org/10.5498/wjp.v2.i5.74.

119. Simpson HB, Foa EB, Liebowitz MR, et al. Cognitive-behavioral therapy vs risperidone for augmenting serotonin reuptake inhibitors in obsessive-compulsive disorder: a randomized clinical trial. *JAMA Psychiatry.* 2013;70(11):1190−1199. Available from: http://dx.doi.org/10.1001/jamapsychiatry.2013.1932.

120. Petretto DR, Preti A, Zuddas C, et al. Study on psychoeducation enhancing results of adherence in patients with schizophrenia (SPERA-S): study protocol for a randomized controlled trial. *Trials.* 2013;14:323. Available from: http://dx.doi.org/10.1186/1745-6215-14-323.

121. Hiemke C, Baumann P, Bergemann N, et al. AGNP consensus guidelines for therapeutic drug monitoring in psychiatry: update 2011. *Pharmacopsychiatry.* 2011;44(6):195−235.

122. Mennickent S, Sobarzo A, Vega M, et al. Determination of clozapine in serum of patients with schizophrenia as a measurement of medication compliance. *Int J Psychiatry Clin Pract.* 2010;14(1):41−46. Available from: http://dx.doi.org/10.3109/13651500903434453.

123. Grundmann M, Kacirova I, Urinovska R. Therapeutic drug monitoring of atypical antipsychotic drugs. *Acta Pharm.* 2014;64(4):387−401. Available from: http://dx.doi.org/10.2478/acph-2014-0036.

124. Miller J, Wehring H, McMahon RP, et al. Urine testing for antipsychotics: a pilot trial for a method to determine detection levels. *Hum Psychopharmacol.* 2015;30(5):350−355. Available from: http://dx.doi.org/10.1002/hup.2482.

125. Zhang G, Terry Jr AV, Bartlett MG. Bioanalytical methods for the determination of antipsychotic drugs. *Biomed Chromatogr.* 2008;22:671−687.

126. Woldu H, Porta G, Goldstein T, et al. Pharmacokinetically and Clinician-Determined Adherence to an Antidepressant Regimen and Clinical Outcome in the TORDIA Trial. *J Am Acad Child Adolesc Psychiatry.* 2011;50(5):490−498. Available from: http://dx.doi.org/10.1016/j.jaac.2011.01.018.

127. Mennickent S, Cifuentes C, Vega M, Rios G, de Diego M. Quantitative determination of paroxetine in human serum by high-performance thin-layer chromatography. *JPC-J Planar Chromatogr Mod TLC.* 2015;28(3):229−233. Available from: http://dx.doi.org/10.1556/1006.2015.28.3.6.

128. Mennickent S, Gonzalez A, Vega M, et al. Development of a validated HPTLC method for quantification of trazodone in human serum. *J Chil Chem Soc.* 2014;59(2):2405−2408.

129. Mennickent S, Fierro R, Vega M, De Diego M, Godoy CG. Quantification of sertraline in human serum by high performance thin layer chromatography as a tool for pharmacotherapy adherence evaluation. *JPC-J Planar Chromatogr Mod TLC.* 2013;26(4):358−362. Available from: http://dx.doi.org/10.1556/JPC.26.2013.4.12.

130. Bosman J, ter Horst PGJ, Smit JP, et al. Adherence of antidepressants during pregnancy: MEMS compared with three other methods. *Ther Adv Psychopharmacol.* 2014;4(2):61−69. Available from: http://dx.doi.org/10.1177/2045125313511486.

131. Bordson SJ, Atayee RS, Ma JD, Best BM. Tricyclic antidepressants: is your patient taking them? Observations on adherence and unreported use using prescriber-reported medication lists and urine drug testing. *Pain Med.* 2014;15(3):355−363. Available from: http://dx.doi.org/10.1111/pme.12300.

132. Chong WW, Aslani P, Chen TF. Effectiveness of interventions to improve antidepressant medication adherence: a systematic review. *Int J Clin Pract.* 2011;65(9):954−975.

133. Rosa AR, Marco M, Fachel JM, Kapczinski F, Stein AT, Barros HM. Correlation between drug treatment adherence and lithium treatment attitudes and knowledge by bipolar patients. *Prog Neuropsychopharmacol Biol Psychiatry.* 2007;31(1):217−224.

134. Mose T, Damkier P, Petersen M, Antonsen S. Therapeutic drug monitoring of lithium: a study of the accuracy and analytical variation between laboratories in Denmark. *Ther Drug Monit.* 2015;37(4):466−471. Available from: http://dx.doi.org/10.1097/FTD.0000000000000198.

135. Medhi B, Prakash O, Jose VM, et al. Seasonal variation in plasma levels of lithium in the Indian population: is there a need to modify the dose? *Singapore Med J.* 2008;49(9):724−727.

136. *National Institute of Mental Health.* Bipolar disorder. < https://www.nimh.nih.gov/health/topics/bipolar-disorder/index.shtml > ; 2015 Accessed 12.01.16.

137. Malhi GS, Tanious M, Bargh D, Das P, Berk M. Safe and effective use of lithium. *Aust Prescr.* 2013;36(1):18−21.

138. Sylvia LG, Reilly-Harrington NA, Leon AC, et al. Medication adherence in a comparative effectiveness trial for bipolar disorder. *Acta Psychiatr Scand.* 2014;129(5):359−365. Available from: http://dx.doi.org/10.1111/acps.12202.

139. de Souza C, Vedana KG, Mercedes BP, Miasso AI. Bipolar disorder and medication: adherence, patients' knowledge and serum monitoring of lithium carbonate. *Rev Lat Am Enfermagem.* 2013;21(2):624−631.

140. *Medimate Minilab.* Medimate MiniLab for lithium analysis in venous serum, venous whole blood and fingerstick whole blood. < https://www.medimate.com/nl/system/files/downloads/Report%201%20Internal%20Lithium%20Validation%20Report%20version%202.0%202014-04-09.pdf > ; 2014 Accessed 15.01.16.

141. *Nottinghamshire Area Prescribing Committee.* Nottinghamshire Health Community Algorithm for Prescribing and Monitoring Lithium within Primary and Secondary Care (Mental Health). < http://www.nottsapc.nhs.uk/attachments/article/3/lithium%20prescribing%20guideline.pdf > ; 2015 Accessed 14.01.16.

142. *World Health Organisation.* Media Centre. Epilepsy. < http://www.who.int/mediacentre/factsheets/fs999/en/# > ; 2015 Accessed 10.12.15.

143. Jones RM, Butler JA, Thomas VA, Peveler RC, Prevett M. Adherence to treatment in patients with epilepsy: associations with seizure control and illness beliefs. *Seizure.* 2006;15(7):504−508.

144. Paschal AM, Rush SE, Sadler T. Factors associated with medication adherence in patients with epilepsy and recommendations for improvement. *Epilepsy Behav.* 2014;31:346−350.

145. Paschal AM, Hawley SR, St Romain T, Ablah E. Measures of adherence to epilepsy treatment: review of present practices and recommendations for future directions. *Epilepsia.* 2008;49(7):1115−1122. Available from: http://dx.doi.org/10.1111/j.1528-1167.2008.01645.x.

146. Mbuba CK, Ngugi AK, Fegan G, et al. Risk factors associated with the epilepsy treatment gap in Kilifi, Kenya: a cross-sectional study. *Lancet Neurol.* 2012;11(8):688−696: doi:10.1016/S1474-4422(12)70155-2.

147. Williams J, Myson V, Steward S, et al. Self-discontinuation of antiepileptic medication in pregnancy: detection by hair analysis. *Epilepsia.* 2002;43(8):824−831.

148. Williams J, Patsalos PN, Mei Z, et al. Relation between dosage of carbamazepine and concentration in hair and plasma samples from a compliant inpatient epileptic population. *Ther Drug Monit.* 2001;23(1):15−20.

149. Malone SA, Eadie MJ, Addison RS, et al. Monitoring salivary lamotrigine concentrations. *J Clin Neurosci.* 2006;13(9):902−907.

150. Wegner I, Edelbroek P, de Haan GJ, Lindhout D, Sander JW. Drug monitoring of lamotrigine and oxcarbazepine combination during pregnancy. *Epilepsia.* 2010;51(12):2500−2502. Available from: http://dx.doi.org/10.1111/j.1528-1167.2010.02771.x.

151. Shah NM, Hawwa AF, Millership JS, et al. Adherence to antiepileptic medicines in children: a multiple-methods assessment involving dried blood spot sampling. *Epilepsia.* 2013;54(6):1020−1027. Available from: http://dx.doi.org/10.1111/epi.12126.

152. Carpentier N, Jonas J, Frismand S, et al. Direct evidence of nonadherence to antiepileptic medication in refractory focal epilepsy. *Epilepsia.* 2013;54(1):e20−e23. Available from: http://dx.doi.org/10.1111/j.1528-1167.2012.03695.x.

153. Hidalgo PRC, Antunes MV, Spaniol B, et al. Adherence to antiepileptic pharmacotherapy: evaluation using a standardized questionnaire and drug concentration in biological samples. *Lat Am J Pharm.* 2014;33(9):1481−1486.

154. Mennickent S, Fierro R, Vega M, De Diego M, Godoy CG. Quantification of lamotrigine in human serum by high performance thin layer chromatography. *JPC-J Planar Chromatogr.* 2011;24(3):222−226. Available from: http://dx.doi.org/10.1556/JPC.24.2011.3.9.

155. Miles MV, Tang PH, Glauser TA, et al. Topiramate concentration in saliva: an alternative to serum monitoring. *Pediatr Neurol.* 2003;29(2):143–147.

156. Patsalos PN, Berry DJ. Therapeutic drug monitoring of antiepileptic drugs by use of saliva. *Ther Drug Monit.* 2013;35(1):4–29. Available from: http://dx.doi.org/10.1097/FTD. 0b013e31827c11e7.

157. *World Health Organisation.* Media Centre. Malaria. Fact sheet No 94 < http://www.who. int/mediacentre/factsheets/fs094/en/ > ; 2015 Accessed 11.12.15.

158. Rahman MM, Dondorp AM, Day NP, et al. Adherence and efficacy of supervised versus non-supervised treatment with artemether/lumefantrine for the treatment of uncomplicated Plasmodium falciparum malaria in Bangladesh: a randomised controlled trial. *Trans R Soc Trop Med Hyg.* 2008;102(9):861–867. Available from: http://dx.doi.org/10.1016/j. trstmh.2008.05.022.

159. Ngasala BE, Malmberg M, Carlsson AM, et al. Effectiveness of artemether-lumefantrine provided by community health workers in under-five children with uncomplicated malaria in rural Tanzania: an open label prospective study. *Malar J.* 2011;10:64. Available from: http://dx.doi.org/10.1186/1475-2875-10-64.

160. Ajayi IO, Browne EN, Bateganya F, et al. Effectiveness of artemisinin-based combination therapy used in the context of home management of malaria: a report from three study sites in sub-Saharan Africa. *Malar J.* 2008;7:190. Available from: http://dx.doi.org/10.1186/1475-2875-7-190.

161. Namusoke F, Ntale M, Wahlgren M, Kironde F, Mirembe F. Validity of self-reported use of sulphadoxine-pyrimethamine intermittent presumptive treatment during pregnancy (IPTp): a cross-sectional study. *Malar J.* 2012;11:310.

162. Bassey G, Nyengidiki TK, John CT. Prevalence of placenta *Plasmodium parasitemia* and pregnancy outcome in asymptomatic patients at delivery in a university teaching hospital in Nigeria. *Niger J Clin Pract.* 2015;18(1):27–32. Available from: http://dx.doi.org/10.4103/1119-3077.146975.

163. Minzi O, Maige S, Sasi P, Ngasala B. Adherence to artemether-lumefantrine drug combination: a rural community experience six years after change of malaria treatment policy in Tanzania. *Malar J.* 2014;13:267. Available from: http://dx.doi.org/10.1186/1475-2875-13-267.

164. Bell DJ, Wootton D, Mukaka M, et al. Measurement of adherence, drug concentrations and the effectiveness of artemether-lumefantrine, chlorproguanil-dapsone or sulphadoxine-pyrimethamine in the treatment of uncomplicated malaria in Malawi. *Malar J.* 2009;8:204. Available from: http://dx.doi.org/10.1186/1475-2875-8-204.

165. Congpuong K, Bualombai P, Banmairuroi V, Na-Bangchang K. Compliance with a three-day course of artesunate-mefloquine combination and baseline anti-malarial treatment in an area of Thailand with highly multidrug resistant falciparum malaria. *Malar J.* 2010;9:43. Available from: http://dx.doi.org/10.1186/1475-2875-9-43.

166. Souares A, Moulin P, Sarrassat S, Carlotti M-P, Lalou R, Le Hesran J-Y. Self-reported data: a major tool to assess compliance with anti-malarial combination therapy among children in Senegal. *Malar J.* 2009;8:257. Available from: http://dx.doi.org/10.1186/1475-2875-8-257.

167. Simba DO, Kakoko D, Tomson G, et al. Adherence to artemether/lumefantrine treatment in children under real-life situations in rural Tanzania. *Trans R Soc Trop Med Hyg.* 2012;106(1):3–9. Available from: http://dx.doi.org/10.1016/j.trstmh.2011.09.006.

168. Cheoymang A, Ruenweerayut R, Muhamad P, Rungsihirunrat K, Na-Bangchang K. Patients' adherence and clinical effectiveness of a 14-day course of primaquine when given with a 3-day chloroquine in patients with *Plasmodium vivax* at the Thai-Myanmar border. *Acta Trop.* 2015;152:151–156. Available from: http://dx.doi.org/10.1016/j.actatropica.2015.08.008.

169. Bruxvoort K, Goodman C, Kachur SP, Schekkenberg D. How patients take malaria treatment: a systematic review of the literature on adherence to antimalarial drugs. *PLoS One*. 2014;9(1):e84555. Available from: http://dx.doi.org/10.1371/journal.pone.0084555.

170. Taneja I, Erukala M, Raju KS, Singh SP, Wahajuddin. Dried blood spots in bioanalysis of antimalarials: relevance and challenges in quantitative assessment of antimalarial drugs. *Bioanalysis*. 2013;5(17):2171–2186. Available from: http://dx.doi.org/10.4155/bio.13.180.

171. Moreso F, Torres IB, Costa-Requena G, Seron D. Nonadherence to immunosuppression: challenges and solution. *Transplant Res Risk Manag*. 2015;7:27–34.

172. Fredericks EM, Dore-Stites D. Adherence to immunosuppressants: how can it be improved in adolescent organ transplant recipients? *Curr Opin Organ Transplant*. 2010;15(5):614–620. Available from: http://dx.doi.org/10.1097/MOT.0b013e32833d3115.

173. Taylor PJ, Tai C-H, Franklin ME, Pillans PI. The current role of liquid chromatography-tandem mass spectrometry in therapeutic drug monitoring of immunosuppressant and antiretroviral drugs. *Clin Biochem*. 2011;44:14–20.

174. Sallustio BC, Noll BD, Morris RG. Comparison of blood sirolimus, tacrolimus and everolimus concentrations measured by LC-MS/MS, HPLC-UV and immunoassay methods. *Clin Biochem*. 2011;44(2–3):231–236. Available from: http://dx.doi.org/10.1016/j.clinbiochem.2010.10.005.

175. Koster RA, Greijdanus B, Alffenaar JW, Touw DJ. Dried blood spot analysis of creatinine with LC-MS/MS in addition to immunosuppressants analysis. *Anal Bioanal Chem*. 2015;407 (6):1585–1594. Available from: http://dx.doi.org/10.1007/s00216-014-8415-2.

176. Koop DR, Bleyle LA, Munar M, Cherala G, Al-Uzri A. Analysis of tacrolimus and creatinine from a single dried blood spot using liquid chromatography tandem mass spectrometry. *J Chromatogr B Analyt Technol Biomed Life Sci*. 2013;926:54–61. Available from: http://dx.doi.org/10.1016/j.jchromb.2013.02.035.

177. Hinchliffe E, Adaway JE, Keevil BG. Simultaneous measurement of cyclosporic A and tacrolimus from dried blood spots by ultra high performance liquid chromatography tandem mass spectrometry. *J Chromatogr B Analyt Technol Biomed Life Sci*. 2012;883–884:102–107. Available from: http://dx.doi.org/10.1016/j.jchromb.2011.05.016.

178. Blanca Martínez Pérez A, López Suárez A, Rodríguez Rodríguez J, Sobrino Márquez JM, Lage Gallé E. Medication adherence in patients who undergo cardiac transplantation. *Transplant Proc*. 2013;45(10):3662–3664. Available from: http://dx.doi.org/10.1016/j.transproceed.2013.11.006.

179. Hsiau M, Fernandez HE, Gjertson D, Ettenger RB, Tsai EW. Monitoring nonadherence and acute rejection with variation in blood immunosuppressant levels in pediatric renal transplantation. *Transplantation*. 2011;92(8):918–922. Available from: http://dx.doi.org/10.1097/TP.0b013e31822dc34f.

180. Stuber ML, Shemesh E, Seacord D, et al. Evaluating non-adherence to immunosuppressant medications in pediatric liver transplant recipients. *Pediatr Transplant*. 2008;12(3):284–288. Available from: http://dx.doi.org/10.1111/j.1399-3046.2008.00923.x.

181. Venkat VL, Nick TG, Wang Y, Bucuvalas JC. An objective measure to identify pediatric liver transplant recipients at risk for late allograft rejection related to non-adherence. *Pediatr Transplant*. 2008;12(1):67–72. Available from: http://dx.doi.org/10.1111/j.1399-3046.2007.00794.x.

182. Pai AL, Rausch J, Tackett A, et al. System for integrated adherence monitoring: real-time non-adherence risk assessment in pediatric kidney transplantation. *Pediatr Transplant*. 2012;16(4):329–334. Available from: http://dx.doi.org/10.1111/j.1399-3046.2012.01657.x.

183. Hugon A, Roustit M, Lehmann A, et al. Influence of intention to adhere, beliefs and satisfaction about medicines on adherence in solid organ transplant recipients. *Transplantation*. 2014;98(2):222–228. Available from: http://dx.doi.org/10.1097/TP.0000000000000221.

184. *World Health Organisation.* Media centre. Asthma. Fact sheet no 307. < http://www.who.int/mediacentre/factsheets/fs094/en/ > ; 2013 Accessed 23.12.15.

185. Bender BG, Bender SE. Patient-identified barriers to asthma treatment adherence: responses to interviews, focus groups, and questionnaires. *Immunol Allergy Clin North Am.* 2005;25 (1):107.

186. Rand CS, Wise RA. Measuring adherence to asthma medication regimens. *Am J Respir Crit Care Med.* 1994;149:S69–S76.

187. Hagan JB, Netzel BC, Matthews MR, Korpi-Steiner NL, Singh RJ. Urinary fluticasone propionate-17beta-carboxylic acid to assess asthma therapy adherence. *Allergy Asthma Proc.* 2012;33(4):e35–e39. Available from: http://dx.doi.org/10.2500/aap.2012.33.3568.

188. Korpi-Steiner NL, Netzel BC, Seegmiller JC, Hagan JB, Singh RJ. Liquid chromatography-tandem mass spectrometry analysis of urinary fluticasone propionate-17beta-carboxylic acid for monitoring compliance with inhaled-fluticasone propionate therapy. *Steroids.* 2010;75(1):77–82. Available from: http://dx.doi.org/10.1016/j.steroids.2009. 10.009.

189. Hagan JB, Taylor RL, Singh RJ. Assessment of synthetic glucocorticoids in asthmatic sputum. *Allergy Rhinol.* 2011;2(1):33–35. Available from: http://dx.doi.org/10.2500/ar.2011.2.0002.

190. Rattenbury JM, Tsanakas J. Acceptance of domiciliary theophylline monitoring using dried blood spots. *Arch Dis Child.* 1988;63(12):1449–1452.

191. Grosso G, Marventano S, Ferranti R, Mistretta A. Pattern of antibiotic use in the community: non-adherence and self-prescription rates in an Italian urban population. *Mol Med Rep.* 2012;5:1305–1310. Available from: http://dx.doi.org/10.3892/mmr.2012.818.

192. *World Health Organisation.* Media Centre. Tuberculosis. Fact sheet no 104. < http://www.who.int/mediacentre/factsheets/fs104/en/ > ; 2015 Accessed 05.01.16.

193. Bachmann LH, Stephens J, Richey CM, Hook III EW. Measured versus self-reported compliance with doxycycline therapy for chlamydia-associated syndromes: high therapeutic success rates despite poor compliance. *Sex Transm Dis.* 1999;26(5):272–278.

194. Yiee JH, Baskin LS, Patel N, Yang J, Disandro M. Prospective blinded laboratory assessment of prophylactic antibiotic compliance in a pediatric outpatient setting. *J Urol.* 2012;187(6):2176–2181. Available from: http://dx.doi.org/10.1016/j.juro.2012.01.130.

195. Al-Ghazawi M, AbuRuz S. Determination of ciprofloxacin in dried blood spots for therapeutic drug monitoring. *Chromatographia.* 2010;71(11–12):999–1005. Available from: http://dx.doi.org/10.1365/s10337-010-1568-z.

196. Czerninski R, Sivan S, Steinberg D, et al. A novel sustained-release clotrimazole varnish for local treatment of oral candidiasis. *Clin Oral Investig.* 2010;14(1):71–78. Available from: http://dx.doi.org/10.1007/s00784-009-0275-3.

197. Rao RN, Vali RM, Ramachandra B, Maurya PK. Rapid determination of rifaximin on dried blood spots by LC-ESI-MS. *Biomed Chromatogr.* 2011;25(11):1201–1207. Available from: http://dx.doi.org/10.1002/bmc.1591.

198. Haidar EA, Lee TD, Barton JD, Coates AR, Mantle PG. HPLC/MS/MS methodology for sensitive quantitation of monic acid A, the metabolic product of the antibiotic mupirocin. *Biomarkers.* 2011;16(5):422–425. Available from: http://dx.doi.org/10.3109/1354750X.2011. 582153.

199. Amlabu V, Mulligan C, Jele N, et al. Isoniazid/acetylisoniazid urine concentrations: markers of adherence to isoniazid preventive therapy in children. *Int J Tuberc Lung Dis.* 2014;18(5):528–530. Available from: http://dx.doi.org/10.5588/ijtld.13.0730.

200. Lee SH, Seo K-A, Lee YM, et al. Low serum concentrations of moxifloxacin, prothiona-mide, and cycloserine on sputum conversion in multi-drug resistant TB. *Yonsei Med J.* 2015;56(4):961−967. Available from: http://dx.doi.org/10.3349/ymj.2015.56.4.961.

201. Tesfahuneygn G, Medhin G, Legesse M. Adherence to anti-tuberculosis treatment and treatment outcomes among tuberculosis patients in Alamata District, northeast Ethiopia. *BMC Res Notes.* 2015;8:503. Available from: http://dx.doi.org/10.1186/s13104-015-1452-x.

202. McCune JS, Reynolds KS. Developing and using therapeutics for emerging infections. *Clin Pharmacol Ther.* 2015;98(4):346−351. Available from: http://dx.doi.org/10.1002/cpt.183.

203. Papaseit E, Marchei E, Mortali C, et al. Development and validation of a liquid chromatography-tandem mass spectrometry assay for hair analysis of atomoxetine and its metabolites: application in clinical practice. *Forensic Sci Int.* 2012;218(1−3):62−67. Available from: http://dx.doi.org/10.1016/j.forsciint.2011.10.012.

204. Gilissen LP, Wong DR, Engels LG, et al. Therapeutic drug monitoring of thiopurine meta-bolites in adult thiopurine tolerant IBD patients on maintenance therapy. *J Crohns Colitis.* 2012;6(6):698−707. Available from: http://dx.doi.org/10.1016/j.crohns.2011.12.003.

205. Gifford AE, Berg AH, Lahiff C, et al. A random urine test can identify patients at risk of mesalamine non-adherence: a prospective study. *Am J Gastroenterol.* 2013;108(2):249−255. Available from: http://dx.doi.org/10.1038/ajg.2012.419.

206. Hawwa AF, AlBawab A, Rooney M, et al. Methotrexate polyglutamates as a potential marker of adherence to long-term therapy in children with juvenile idiopathic arthritis and juvenile dermatomyositis: an observational, cross-sectional study. *Arthritis Res Ther.* 2015;17:295. Available from: http://dx.doi.org/10.1186/s13075-015-0814-z.

207. Reinders MK, Nijdam LC, van Roon EN, et al. A simple method for quantification of allopurinol and oxipurinol in human serum by high-performance liquid chromatography with UV-detection. *J Pharm Biomed Anal.* 2007;45(2):312−317.

CHAPTER 5

Medication Adherence: Where Do We Go From Here?

INTRODUCTION

Previous chapters in this book have coincidentally examined several of the pointers in the 2010 National Institutes of Health report,[1] namely:

- The development of miniaturized devices to allow patients to take a greater role in their own healthcare which will become more personalized
- Point-of-care or bedside monitoring using small blood spot samples will be possible
- Low cost point-of-care devices can be used for the disadvantaged and underserved population of the United States as well as in the developing world
- New monitoring methods would be developed particularly for cancer patients

We concluded that there was recognition, from around the world, of the value of reliable and affordable direct monitoring methods especially for the routine assessment of medication adherence. The information provided by monitoring of this type empowers clinicians to make decisions that will significantly improve individual patient healthcare. It will also impact on other stakeholders such as healthcare systems and pharmaceutical companies. Healthcare costs are rising dramatically as populations around the world get older and require treatment for multiple ailments and as improved healthcare reaches even greater numbers of people with more and better services. Thanks to the World Wide Web, people are becoming more interested in health-related problems and are demanding more information and services from their healthcare providers.

IDENTIFYING PERCEIVED CHALLENGES

The potential to provide medication adherence assessment tests for a wide range of illnesses has been demonstrated in Chapters 3 and 4

Analytical Chemistry for Assessing Medication Adherence. DOI: http://dx.doi.org/10.1016/B978-0-12-805463-5.00005-5

but to develop such monitoring services raises questions around the selection of the analytical technique, the capacity, and the financial ability to deliver them. While immunoassay provides rugged "kit-based" analyses which are just good enough for the current needs of clinicians,[1] there are areas of poor specificity and no opportunity for multiple data points from a single sample. These limitations promote the LC-MS/MS systems as the currently "best technique" but there are several perceived challenges to the use of mass spectrometry in clinical laboratories.[2-5] An appreciation of these problems and some discussion of possible routes forward form the central theme of this chapter which also includes some suggested alternative approaches to the pivotal problem of objective adherence monitoring.

The perceived challenges to the wider use of LC-MS/MS techniques include:

• shortage of suitably trained staff
• complexity and cost of the equipment
• software and data handling
• limitations on sample throughput
• lack of dialogue between clinicians and mass spectrometer providers
• instrument manufacturers' concerns
• regulatory actions

These are discussed below.

Shortage of Suitably Trained Staff

As the number of elderly people needing some form of medical care increases so does the likely demand on clinical test facilities and staff. In the United Kingdom, for example, the number of people aged 60 years and older is projected to increase by at least 25% over the next 15 years.[6] This is a slower rate than for the rest of the world. Moves have already been made to automate some systems within clinical laboratories to reduce anticipated demands for additional staff. Many clinical laboratories use commercially prepared reagents, standard solutions, and QC processes to save time, but now technical staff have become less familiar at calculating and weighing out materials in preparation for tests.[3] Complete kits for LC-MS analyses including the provision of the column and mobile phases is now a reality. Even with this provision, liquid chromatography and mass spectrometry are both seen as high−performance, high-maintenance techniques needing skilled staff.[2,4] Indeed some institutions cite lack of skilled HPLC staff for not developing new analytical methods.[4] To develop an adequate expertise with mass spectrometry requires several years of hands-on experience[2,5] representing a skills gap rather than a knowledge gap. There may also be a shortage of staff qualified to deal with the high volume of data output both in terms of straight information technology and data handling and also explaining the meaning and value of the data to the clinicians.[2] Specialized in-house training from the MS manufactures can overcome the problem in the short term until the more automated push-button analyzer is developed.

Complexity and Cost of the Equipment

There can be no doubt that mass spectrometers are complex instruments, but over the last 30 years manufacturers have reduced a room full of stainless steel tubes and massive magnets to bench-top units with superior performance. This has been achieved by developing new technologies, for example the Orbitrap; improving older systems, for example the time of flight (ToF) systems and reducing the bulk of the vacuum pumps and other equipment needed for the MS systems. Having achieved this small footprint either a gas chromatograph or a liquid chromatograph is added to the front end. For clinical samples, for example, some form of sample handling and/or clean-up process may be added (see Fig. 5.1) and interconnectivity begins to suffer. System robustness also suffers with increasing numbers of additional nonoriginal equipment manufacturer items that are added. Furthermore, the service laboratory information system may no longer be able to track the samples with all these add-ons to the MS system. The general call from the clinicians[1−5] seems to be for a fully integrated push-button system that will receive a raw sample, pre-prepare as necessary, carry out specified analyses, and deliver reports compatible with the laboratory central information system as exemplified by Fig. 5.1. As Jannetto and Fitzgerald suggested, not until the MS becomes simply a detector on an automated FDA-approved platform will its full benefit be realized.[3]

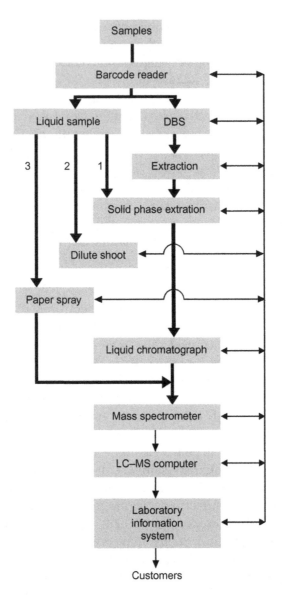

Figure 5.1 Schematic diagram of a proposed fully integrated clinical sample analysis for either a dried blood spot (DBS) or a liquid biosample. The possible routes (thick lines) are determined by the barcode via the laboratory information system (thin lines). Selection of routes 1, 2, or 3 depends on the nature of the liquid biosample.

Which MS system to select presents another series of questions (see Chapter 3), but for routine analyses the triple quadrupole tandem MS (LC-MS/MS) system seems to be the system currently favored based on the number of investigations using this instrument reported in

Chapter 4. The choice of other instruments may depend on the nature of the work whether it be high resolution (Orbitrap) or high mass range (ToF).

The arguments needed to justify the costs of these systems will inevitably depend on the local situation and whether the instrument is meant simply to provide a service or to do so and make a profit. The arguments will include:

- The ability to do new analytical tests providing better patient care
- The ability to do multiple determinations on a single biosample
- Do current analytical tests more efficiently thereby giving cost savings
- Add other test profiles to the instrument, for example, new-born screening
- Remove the need to send samples out for costly external testing

Both the capital cost and the running costs of the analytical instrument must be determined. Consumables, especially the solvents and columns for the LC-MS/MS, can be relatively expensive as can service contracts on all the separate components of the overall system. Currently, this usually means dealing with different suppliers with plenty of room for differences of opinion when instrument problems develop.

Software and Data Handling

In several instances, software and data handling were cited as causes for concern.[3,5] In the author's experience of several MS systems, the software appears to be written more to demonstrate the skill of the writer rather than to make the analytical process as intuitive as possible. This may lead to longer training time for new staff than might otherwise be required. This problem may increase as more multiplexing of data collection from single samples is required. The benefits of simplifying the MS approach and gaining FDA approval is evident from the advances in productivity and time savings made when using MALDI-time of flight (MALDI-ToF) systems for the routine clinical identification of microbes.[7]

Data processing can also be a problem when quantitative data is required from many target compounds from a single sample. The data capture, processing, and reporting capability must be equal to the task of dealing with high-throughput systems without delays in sample processing.[5]

Limitations on Sample Throughput

The minimum time unit within an LC-MS/MS instrument is the time required to complete either one complete m/z scan or one multiple reaction monitoring (MRM) determination (see Chapter 3). Decreasing these time periods is important because they affect the number of data points that can be obtained during the elution time of a compound from the chromatographic separation and into the MS analyzer. Both of these factors are under constant review by the manufacturers who are aware of a possible trade-off between scan speed and detection capability. Any separation system at the front end of an MS will have periods when no compounds of interest are eluting and while efforts to reduce elution times are common this problem cannot be avoided. Idle MS time waiting for target compounds can be reduced by interfacing several LC systems to the input of a single MS via a multiposition switching valve.[3,8] The timing of injections and switching the different eluents into the MS will depend on both the retention time of the compounds of interest and the reequilibration time for the column.

Sample preparation prior to analysis may also present restrictions on throughput although some automated process for both liquid samples and DBSs have been discussed in Chapter 2. There is an opportunity here for the development of "smart sample collection" devices which will pre-process the sample prior to reaching the clinic. The conventional DBS card is a simple example of this approach as the unwanted components of blood are immobilized on the cellulose and the target analytes can be extracted by an automated flow through process with no punching of discs required. An alternative approach is taken in the Noviplex card where a plasma sample is produced from whole blood by in-card filtration. Both of these processes effectively move sample preparation time away from the clinical laboratory toward the patient.

Potential elimination of the time-consuming steps of sample preparation and chromatographic separation using PaperSpray ambient ionization with transfer of the ions directly into the MS have been reported by Conaway.[9] The advantages of this approach include no sample preparation, no chromatographic system to maintain, minimum solvent use, and reduced training. The disadvantages are slightly poorer detection limits and no chromatographic separation, specificity depending solely on MRM or HRMS capabilities. This technique and other ambient ionization techniques used to monitor therapeutic drugs in DBS, plasma, oral fluids, or urine have recently been reviewed by Ferreira et al.[10] and have potential for medication adherence monitoring.

A market survey carried out in 2014 among research laboratory personnel mainly in pharmaceutical and biotechnical industries and academia concluded that between 10−25 k samples processed/24-hour day constituted real high throughput.[5] This would limit each sample analysis time to less than 10 seconds which would seem ambitious in a clinical setting.

Lack of Dialogue Between Clinicians and Manufacturers

It is relatively easy from an academic viewpoint to support a much greater rate of information exchange and discussion but the need to protect intellectual property is also recognized. There does however have to be an exchange of information in order to move development forward toward a sustainable market for the analytical equipment. One approach is the development of "wide audience" conferences such as Mass Spectrometry Applications to Clinical Laboratories (MSACL)[11] where the target audience includes pathologists, clinical laboratory directors and scientists, mass spectrometry and automated sample processing providers, as well as laboratory diagnostic providers. Meetings of this type will also be beneficial to healthcare administrators, laboratory quality control, and standards scientists. MSACL meetings are held in both the United States and Europe. The International Association of Therapeutic Drug Monitoring and Clinical Toxicology (IATDMCT)[12] organizes even "wider audience" meetings with representatives from immunoassay manufacturers present. Both groups are ideal forums to discuss direct assessment methods of adherence to medication. Another approach in this area might be through the American Association of Clinical Chemistry who organize meetings and webcasts on the clinical applications of MS.[13]

Instrument Manufacturers' Concerns

New developments in MS systems usually only occur in a stepwise manner and companies are therefore keen to protect and exploit their intellectual property rights. This means different manufacturers can offer new developments in ionization, detection, overall sensitivity, or number of samples handled at different times as their own internal research programs come to fruition. There appears to be a lack of clarity from the end-user clinical fraternity as to exactly what form of instrument is required; standalone laboratory-based or mobile bedside capable? Most importantly, some indication of the potential market and particularly the nature of the system "front end" sample handling requirements would help investment decisions by the manufacturers. A knowledge of the type or types of biosample to be analyzed would define the nature of the additional "front end" sample preparation instrumentation which needs to be integrated into the MS system.[4] Alongside all of these uncertainties, clear guidance from a regulatory perspective would be welcome in order to inform critical strategic business activities that will advance science while meeting all regulatory requirements worldwide.[4]

Regulatory Actions

Until recently, there was little guidance applicable to the use of LC-MS systems for bioanalysis in clinical laboratories. Laboratories chose to validate their LC-MS methods by following other industry standards such as the FDA Guidance for Industry: Bioanalytical Method Validation,[14] the European Medicines Agency Guidelines on Bioanalytical Method Validation,[15] or the Scientific Working Group for Forensic Toxicology Standard Practices for Method Validation.[16] Currently, there is only one quantitative test approved by the FDA all others fall under the laboratory developed test (LDT) classification.[17] A new standard for clinical mass spectrometry CLSI C62-A has been produced by the Clinical and Laboratory Standards Institute and while many of the guidelines parallel those in the documents cited above CLSI C62-A provides a more comprehensive approach to the evaluation of trueness, carryover, linearity, and interferences. It is interesting to note that this document is specific to systems relying on MRM (or SRM) methods as the primary mode of analyte identification. Other techniques operating in full-scan mode and relying on library searching for identification are in need of similar guidance.[17]

WAYS FORWARD

Evidence-Based Treatment and Personalized Medicine

Clinicians need measures to assess the effects of prescribed drug therapy to help with the clinical decision-making process. The question of adherence to medication is an integral part of the determination of the optimum level of the appropriate drug(s) for an individual patient based on their biosample drug concentration data. This would be dependent on drug pharmacokinetics and patient pharmacogenetics. In the event of poor patient progress the data available would indicate either poor adherence or adequate adherence but poor choice of a drug or an incorrect diagnosis. The quantification of prescribed drug levels in patient biosamples, compared with its therapeutic window,[18] would provide clinicians a solid evidence-based means for further treatment. This would therefore be a step change in the way clinicians support their patients.

In the immediate future even if such analyses are carried out with current analytical equipment, it will provide a beneficial detectable improvement in patient care which will be the spur for the more rapid development of a sophisticated clinical analyzer. The current situation is the patient goes to the clinic to provide the sample which is then taken to be analyzed. A small number of clinics maximizes the number of samples per analyzer but inconveniences patients who are forced to travel some distance. Home sample collection removes this objection and could generate further samples relating to other diseases, for example sexually transmitted diseases.[18] This scenario suggests a multipurpose central analyzer or analyzers in the near future with full miniaturization to a mobile instrument in the longer term.

The patient can become involved in their healthcare process when the optimization of the medication prescription based on their biosample drug concentration data is explained to them by their clinician, pharmacist, or other healthcare professional. This concept of a "personalized for you" medication care scheme could become routine and lead to long-term savings for the healthcare provider by improving patient outcomes, optimizing medicine usage, and eliminating unnecessary hospital readmissions and untimely patient deaths.

Home Sampling

A major step forward in objectively assessing medication adherence would be to provide an individual patient with the capability to collect their own biosample with subsequent transport for analysis, thus eliminating the need for a phlebotomy appointment or clinic visit. The

DBS-based sampling devices, highlighted in Chapter 3 lend themselves ideally for such home sampling by a patient. With a global burgeoning elderly population in which multiple chronic conditions and poly-pharmacy prescribing are common, and for whom attendance at a phlebotomy clinic adds a further layer of complexity, the home-sampled DBS approach may provide healthcare providers an innovative way of providing evidence-based drug treatment, thereby shifting the balance of healthcare provision from hospitals to the community.[19] The vision is for a paradigm shift in the medication adherence investigations of various clinical conditions involving blood sampling away from hospital-centered activity toward patient-focused home sampling in order to widen access to healthcare. Such sampling strategies could lead to increased patient benefit by offering multiple blood sampling opportunities at home or in a care home with the aim of reducing hospital visits, for example, following postacute treatment for a cardio-vascular event. Of primary importance in such investigations is the ability of the patients to collect acceptable biosamples on their own away from a clinic. To facilitate this, it must be easy for the patients to handle the collection device which will take a fixed volume sample and be protected against accidental contamination. Returned conventional DBS cards, for example, must be checked to ensure adequately sized spots which are not smudged or misshapen. Such spots would be rejected since it would not be possible to obtain meaningful quantitative data from their analyses. The performance of other DBS sampling devices under true home patient self-sampling conditions have yet to be reported. From the service providers point of view the cost of the sampling device will be important and this will be balanced by any increased level of samples returned and sample security. There may also be concerns about the level of sample preparation which may be accomplished in the sampler.

Future Clinical Practice Guidelines
Future international clinical practice guidelines with recommendations for objective direct measures would provide the impetus and support for this approach to be more readily adopted to address the problem of poor medication adherence.

Microfluidics
Objective medication adherence monitoring studies require information regarding drug concentration in a biosample at time points during the process of absorption, distribution, metabolism, and excretion. In

order to obtain reproducible and good quality data, the sampling method is as important as the bioanalytical method and many of the "gold standard" methods rely on multistep sample preparation and analysis techniques. The field of microfluidics which is emerging as a significant trend in bioanalytical instrumentation could address these pertinent issues. Microfluidics or "lab-on-a-chip" is generally used to describe any technology that moves microscopic and nanoscale volumes of liquid through microsized channels on a microelectromechanical system.[20] Advances have been made in the miniaturization and automation of microfluidic devices with features including the ability to process very small sample volumes (pL to nL) in short analysis times (minutes or seconds), amenable to integration/multiplexing and high-throughput analysis, automation, small footprint, cost, disposability, and portability.[21] There are some examples of commercialized microfluidic laboratory based devices based on paper or polymeric materials for healthcare applications. Namely, Abbott's iSTAT device, Epocal's Flexcard, or Abaxis' compact-disc blood analyzers which perform complex assays on 50−100 μL volumes of blood. Multiplexed formats are used for the detection and quantification of electrolytes, gas, small molecules, or protein markers of metabolic, liver, renal, or cardiac conditions.[21] These technologies have enormous potential for cost-effective tests for medication adherence monitoring purposes provided that they meet requirements of analytical specificity, sensitivity, and detection limit as well as overcoming national and international regulatory barriers. Ease of use and reliability would be other important requirements. Point-of-care diagnostic tests for home use have also been designed based on microfluidics, for the semiquantitative assessment for the presence of a biochemical species of interest (eg, blood glucose monitoring). If this initiative is extended to medication adherence monitoring, then it could promote a shift away from traditional clinical laboratory testing to near-patient settings, thus providing clinicians or pharmacists with timely information about medication-taking behavior so as to make informed clinical decisions regarding treatment. In this setting, there would be no requirement for biosamples to be sent to a laboratory for analysis. In such a setting the clinician or healthcare professional would initiate the adherence monitoring test and receive the results on the spot, thereby saving time for providing results rapidly. These initiatives are in line with the National Institute of Health's 2010 report to effect a shift toward personalized healthcare[1] and would also need to address the varied healthcare needs worldwide as highlighted in Chapter 4. It is interesting to speculate if the combination of:

A blood spot + "Lab-on-a-chip" + Mobile phone = Future of bioanalysis

Perhaps the first move in this direction is the volume measured DBS microfluidic sampling system using capillary forces and dissolvable film valving to provide a timing function.[22] The volume metering DBS chip consists of a capillary tube mounted on a dissolvable plastic film mounted on a piece of DBS card. A drop of blood from a fingerprick is deposited at one end of the capillary tube and will start to fill the capillary and at the same time it will dissolve the plastic layer on the DBS card. The blood, in the capillary tube, reaches the far end of the tube and will again start to dissolve the plastic layer on the DBS card. Excess blood from the initial drop is transferred on to the card when the first area of plastic ruptures and a measured volume of blood, in the capillary tube, is collected on the card when the second plastic film ruptures. This approach provides a fixed volume of blood independent of the haematocrit level.

CONCLUSIONS

Adherence to prescribed drug therapy is a means of achieving therapeutic levels of drug in the blood. Any direct assessment of these levels has to be the best way of benefiting the clinicians, the patients, and the healthcare systems globally. In the immediate future the results of such a test will provide the evidence base to aid the clinical decision-making process and to maximize patient benefit from the prescribed drug therapies. Home sampling will ease the burden on overstretched healthcare facilities globally and will augment this process.

Effective adherence to prescribed pharmacotherapy, for an individual, will be confirmed if the measured blood drug levels are within the therapeutic windows. Thus the analysis of biosamples will facilitate the development of personalized medicine to the benefit of patients globally. Advances in analytical instrumentation will mean that this will no longer be a dream but a reality in the long term. None of this will be possible without the global use of analytical chemistry!

REFERENCES

1. *NIH.* Point of care diagnostic testing. <https://report.nih.gov/nihfactsheets/ViewFactSheet. aspx?csid=112>; 2010 Accessed 1.2.16.

2. *Technology networks—The online scientific community—News.* Education and expense: the barriers to mass spectrometry in clinical laboratories? <http://www.technologynetworks.com/ news.aspx?id=184032>; 2015 Accessed 25.11.15.

3. Jannetto PJ, Fitzgerald RL. Effective use of mass spectrometry in the clinical laboratory. *Clin Chem.* 2016;62(1):92–98.

4. Bachmann L, Hanash S, Hart BA, et al. A spectrum of views on clinical mass spectrometry. Annesley T. Diamandis E (mods). *Clin Chem.* 2016;62(1):30–36.

5. *Chromatography—The online scientific community—News.* High throughput mass spectrometry-based screening assay trends. <http://www.technologynetworks.com/ Chromatography/news.aspx?ID=185391>; 2015 Accessed 25.11.15.

6. *Age UK.* Later life in the United Kingdom. <http://www.ageuk.org.uk/Documents/EN-GB/ Factsheets/Later_Life_UK_factsheet.pdf?dtrk=true>; 2016 Accessed 25.1.16.

7. Clarke AE, Kaleta EJ, Arora A, et al. Matrix assisted laser desorption ionisation-time of flight mass spectrometry: a fundamental shift in the routine practice of clinical microbiology. *Clin Microbiol Rev.* 2013;26:547–603.

8. Bayliss MK, Little D, Mallett DN, et al. Parallel ultra high flow-rate liquid chromatography with mass spectrometric detection using a multiplex electrospray source for direct, sensitive determination of pharmaceuticals in plasma at extremely high throughput. *Rapid Commun Mass Spectrom.* 2000;14:2039–2045.

9. Conaway MP. Rapid biofluid analysis using the Prosolia Velox 360™ PaperSpray^R system and mass spectrometry. <https://www.thermoscientific.jp/content/dam/tfs/ ATG/CMD/cmd-documents/sci-res/pres/ms/lc-ms/acc/SP-64632-LC-MS-Prosolia-Velox-360-PaperSpray-SP64632-EN.pdf>; 2013 Accessed 1.2.16.

10. Ferreira CR, Yannell KE, Jarmusch AK, et al. Ambient ionisation mass spectrometry for point of care diagnostics and other clinical measurements. *Clin Chem.* 2016;62(1):99–110.

11. *Mass Spectrometry Applications to Clinical Laboratories (MSACL).* Accelerating clinical mass spectrometry. <https://www.msacl.org/index.php?header=Learning>; 2015 Accessed 2.2.16.

12. *The International Association of Therapeutic Drug Monitoring and Clinical Toxicology.* <https://www.iatdmct.org/events/iatdmct-congress.html>; 2016 Accessed 2.2.16.

13. *American Association for Clinical Chemistry.* <https://www.aacc.org/search-results#q/ q=mass%20spec>; 2016 Accessed 2.2.16.

14. *U.S. Department of Health and Human Services, Food and Drug Administration, Center for Drug Evaluation and Research.* Guidance for industry: bioanalytical method validation. <http://www.fda.gov/downloads/Drugs/Guidances/ucm070107.pdf>; 2001 Accessed 1.11.15.

15. *European Medicines Agency.* Guideline on bioanalytical method validation. <http://www.ema. europa.eu/docs/en_GB/document_library/Scientific_guideline/2011/08/WC500109686.pdf>; 2011 Accessed 1.11. 2015.

16. *Scientific Working Group for Forensic Toxicology.* Standard practices for method validation in forensic toxicology. <http://www.swgtox.org/documents/Validation3.pdf>; 2013 Accessed 1.2.16.

17. Lynch KL. CLSI C62-A: a new standard for clinical mass spectrometry. *Clin Chem.* 2016;62 (1):24–29. Available from: http://dx.doi.org/10.1373/clinchem.2015.238626.

18. Tanna S, Lawson G. Dried blood spot analysis to assess medication adherence and to inform personalization of treatment. *Bioanalysis*. 2014;6(21):2825–2838.

19. Tanna S, Lawson G. Editorial—self-sampling and quantitative analysis of DBS: can it shift the balance in over-burdened healthcare systems? *Bioanalysis*. 2015;7(16):1963–1966.

20. Gomez FA. Bioanalytical application in microfluidics. *Bioanalysis*. 2010;2(10):1661–1662.

21. Lazar IM. Microfluidic devices in diagnostics: what does the future hold? *Bioanalysis*. 2015;7 (20):2677–2680.

22. Lenk G, Sandkvist S, Pohanka A, et al. A disposable sampling device to collect volume-measured DBS directly from a fingerprick onto DBS paper. *Bioanalysis*. 2015;7(16):2085–2094.

Printed in the United States
By Bookmasters